Agri-Innovations and Development Challenges

Innovation in Engineering and Technology Set

coordinated by
Dimitri Uzunidis

Volume 8

Agri-Innovations and Development Challenges

Engineering, Value Chains and Socio-economic Models

Edited by

Vanessa Casadella
Dimitri Uzunidis

iSTE

WILEY

First published 2023 in Great Britain and the United States by ISTE Ltd and John Wiley & Sons, Inc.

ISTE Ltd
27-37 St George's Road
London SW19 4EU
UK

www.iste.co.uk

John Wiley & Sons, Inc.
111 River Street
Hoboken, NJ 07030
USA

www.wiley.com

Library of Congress Control Number: 2023936092

British Library Cataloguing-in-Publication Data
A CIP record for this book is available from the British Library
ISBN 978-1-78630-915-0

Contents

Introduction

Agri-Innovations in Developing Countries: Relevant Responses to the Environmental, Food, and Social Crisis

I.1. Introduction

A total of 828 million people were suffering from food poverty in 2021, representing 9.8% of the world's population. In 2019 and 2020, this share was 8% and 9.3%, respectively. Nearly 2.3 billion people (29.3% of the world population) were moderately or severely food insecure. In 2021, 31.9% of the world's women were moderately or severely food insecure, compared to 27.6% of men. Note that 3.1 billion people could not afford a healthy diet. In most countries of sub-Saharan Africa and South Asia, as well as in some Central American countries, the food emergency is absolute. In the world, due to the increase in social inequalities, conflicts and especially climate change, food is far from being guaranteed and the conditions of agricultural production are worsening. The latest FAO report (2022) is more than alarming. If nothing is done to curb global warming, improve the sharing of wealth and stop military conflicts between nations, it is the world's food security that will be threatened: loss of nutrients in staple foods, reduced harvests, large-scale climatic disasters in several regions, etc. The risks are high and the first signs of this agricultural and food crisis are already visible. Food security will be increasingly compromised by future climate change due to lower yields, especially in tropical regions, higher prices, reduced quality of nutrients and disruptions in the supply chain.

Introduction written by Dimitri UZUNIDIS and Vanessa CASADELLA.

Ecologically responsible agricultural production, less harmful to the environment and biodiversity, requires targeted innovations through research processes, dissemination of relevant knowledge, for the experimentation of this knowledge in the field, productive adaptation, integration in crops, livestock and processing or by promoting adapted commercial channels and creative entrepreneurship.

I.2. Food security, climate change and socioeconomic disorder

Food safety can only be understood holistically: it encompasses all elements that can impact food safety at each stage of the agri-food value chain. Applying this approach means that food safety is not only about the development of the final product that starts with primary production, with it being processed in the factory and sold in the market, but also emphasizes the need for interaction between all stakeholders in the value chain and between these and the characteristics of their environment, such as climate change, demographics, the state of the economy, social structure, etc. (Gaitis and Ouzounidou 2017).

The correlation between food security and climate change is part of a sociotechnical and economic context based on the following three pillars (Uzunidis 2019):

a) The rapid increase in the world population associated with malnutrition, poverty and famine: the FAO estimates that by 2050, the global demand for food and feed will increase by 60%. Two billion people in the world are overweight; one billion are chronically undernourished.

b) The domination of the petrochemical paradigm in economic activity.

c) The socioeconomic organization and the global model of governance: industrial and intensive agriculture and breeding, *transnational oligopoly* (Big Food), conflicts of sanitary norms and patents, protectionist tendencies, unequal distribution of income and access to food resources, decrease or sudden fluctuations of raw material prices, etc. – "planet for profit strategies and policies" or "profit strategies against planet and people". The current food system generates one-third of greenhouse gas emissions!

The correlation between agriculture, food security and climate change can be seen in the definitions of these two concepts: food security is achieved when all people, at all times, have physical and economic access to sufficient, safe and nutritious food to meet their energy needs and food preferences for an active and healthy life.

The following four conditions must be met to achieve food safety:

– Food availability: this first condition refers to the quantity of food available that is sufficiently abundant to satisfy the vital needs of the population. The growth of agricultural production must therefore be positively correlated with the increase in population.

– Physical, social and economic access to food: this second condition refers to demand. The evolution of the supply of agricultural products must be correlated with the evolution of demand, which, correlated with the level of income, must be satisfied to cover vital individual and collective needs.

– The quality of the food consumed: this condition refers to the nature of the products consumed and must be correlated with other parameters related to health, the physical environment and the individual and collective socialization of the populations.

– Stability or sustainability, in the sense that the food security sought is not a one-time event and must be part of a sustainable and socially equitable development perspective.

Food security is therefore a complex, multidimensional concept that depends on agricultural and cooperation policies, economic choices, trade regulations and social policies and standards.

Climate change refers to all variations in climatic characteristics. Air pollution, resulting from human activities, modifies the climate, in the sense of a global warming. This phenomenon can cause significant damage: rising sea levels, increased extreme weather events (droughts, floods, cyclones, etc.), destabilization of forests, threats to freshwater resources, agricultural difficulties, desertification, reduction of biodiversity, extension of tropical diseases, etc.

The effects of climate change are increasingly felt in agricultural production. Inaction in decision-making to adapt to this change is becoming an insurmountable constraint, with the risk of leading to a total impasse. In any case, adjustments must be multi-level and implemented at the international, national and local levels on the basis of a set of precise and imperative actions to address the issues raised by climate change. Climate change is a major concern in primary production, marketing and processing, in particular, of plants, as extreme weather events in recent decades have resulted in lower crop yields, loss of production and income due to harsh bioeconomic conditions, and a decrease in the quantities of nutritious quality products packaged and processed (Ortiz-Bobea et al. 2021).

The challenge, according to the experts (Christopoulos and Ozounidou 2021), is to articulate an economic approach, environmental protection and new models of technological, organizational and social innovation and social justice in a context of economic globalization, privatization/commodification of natural resources and deepening inequalities. Indeed, regarding the variations and the extreme climatic conditions:

a) They are negatively impact the availability, access, use and stability of agricultural and food resources, as well as feeding, care and health practices. Direct and indirect climate impacts have a cumulative effect, leading to increased food insecurity and malnutrition.

b) They are negatively correlated with agricultural productivity, food production, cropping patterns, thus contributing to food shortages, and nutrition: the nutritional quality and diversity of food produced and consumed deteriorates, and water stress increases, resulting in increased health risks, diseases and poverty.

c) They lead to sharp increases and volatility in primary product and food prices, often associated with losses of farm income reducing access to food and negatively affecting the quantity, quality and dietary diversity of food consumed.

I.3. Paradigm shift

As climate change contributes to biodiversity loss, which poses an imminent and ongoing threat to food security and livelihoods, the urgency of the paradigm shift is equally clear (Schlaile et al. 2022). The food chain faces increasing threats each year from repeated droughts, floods, wildfires, loss of wildlife and the emergence of new pests, microbes and viruses. In response to these threats to cope with climate change, there is a need to (a) implement measures to plan new plantings by selecting appropriate varieties according to local climatic conditions, (b) promote research and development (R&D) to discover/design stress-resistant varieties, (c) record climatic parameters, (d) provide timely information to growers on the importance of modifying conventional farming practices (e.g. use of chemicals and intensive farming), (e) use modern technological tools (precision or smart farming), (f) develop irrigation and flood protection infrastructure, (g) restructure some crops to higher elevations where conditions are more suitable, (h) provide artificial soil cover crops and (i) rely on traditional and endogenous knowledge and models of community self-organization that can be compatible with “doing” and “sustaining”. In animal husbandry, the use of on-farm technologies to improve herd management for maximum production results at the lowest possible cost is another avenue to explore. Finally, and for all this, water management is the key issue. The achievement of all these objectives, which are difficult to reach but urgent, requires the introduction of major innovations in the productive systems.

To cope with extreme weather events, satisfy ever-changing consumer tastes, and meet ever-increasing demand for high-quality food, or simply subsistence products, farmers today and in the decades to come are called upon to grow more crops using less water and inputs, adopt new technologies and invest in improving the quality of their crops. The need for cooperation in the field of agricultural technology is imperative for all of the above because more must be produced, qualitatively and quantitatively, with fewer natural resources and inputs: *more with less*. And in this endeavor, no institution, idea or technology can be left out.

In short, five main pathways are proposed to ensure the reduction of climate change impacts on food security:

1) control of food production by the producer and the consumer – this leads to the rationalization of the agricultural activity in relation to the available resources and change in the "Big Food" paradigm;

2) developing ecological agriculture: change in the petrochemical paradigm;

3) promoting biodiversity;

4) strengthening the resilience of food systems;

5) promoting biochemistry and the bioeconomy in general.

It is a question of re-establishing a more balanced relationship between the biosphere and the socioeconomic sphere. The bioeconomy associated with digital technologies and frugality calls for the design and application of new sociotechnical models. The implementation of ecodevelopment strategies based on the creation of a development model linked to the needs of populations, the participatory and contributive process and the characteristics and resources of the territories is not without consequences for the agricultural world, which must both feed the population and provide sufficient organic resources to meet our energy needs. Thus, the diversity of environments, contexts, modes of coordination and the influence of existing institutions must be taken into account to ensure structural changes (Debref et al. 2022).

As the authors of this book show, it is a question of designing production systems according to the functionalities offered by ecosystems: reduction of pressure on the environment and preservation of natural resources. The challenge is to articulate the economic approach with environmental protection and social justice (equity in the distribution of wealth, pacification) in a context of economic globalization, commercialization of natural resources and increasing inequalities.

The paradigm shift (in the sense of a new mode of production, consumption, distribution and socialization) is a burning issue! However, the ecological transition for food security comes up against insufficient efforts to reduce pollution on a global scale, as well as the absence of a global governance of common goods and resources. On the other hand, strategies for appropriating life by "Big Food", lobbying and standardization by anticipating the use of alternative technologies slow down the diffusion of green innovations, increase their cost of design and application, prohibit their use on a large scale and prevent design thinking in this field from blossoming on a new trajectory of creativity and project engineering.

I.4. Capacities and potential for agro-innovations in the South

The environmental crisis is all the more perceptible in developing countries, as their agricultural and agri-food production capacities are weak. In most countries in Africa, Latin America and South Asia, citizen mobilization (farmers, consumers, associations, small businesses, etc.) is taking shape through small-scale actions, but they are bringing about innovations that are sometimes frugal, sometimes on a larger scale. Value chains are taking shape, developing and asserting themselves in their economies on both local and global levels.

The notions of innovation systems and capacities have been used both by public policies and in the academic sphere to relate the innovation performance of territories, sectors, regions and nations. Innovation systems are represented by the connection between actors from the academic, political and productive worlds to organize, mobilize and valorize material, financial and human resources for the purpose of innovation. Innovation capabilities are defined both by the knowledge, skills and techniques available to an economy, and by the interactions between innovation actors in that economy to explore and exploit opportunities to develop new products or services based on the needs of individual or collective consumers or users (Casadella et al. 2015; Casadella and Uzunidis 2017).

In order for innovation capacities to become a system, it is necessary to multiply, fluidify and strengthen interactions between private and public innovation actors through the formation of value chains. The ability to innovate and adapt to economic and socioecological hazards thus depends on institutional arrangements in addition to technological development (Asayehegn et al. 2017). A value chain is constructed by three sets of elements that characterize the systemic nature of the relationships between the actors that form it: (a) the succession of integrated transformation operations that are separable and identifiable, but that are linked to each other by technological sequences that ensure their integration; (b) the commercial and financial relationships between all stages of production and between customers and suppliers that ensure the coordination of the activities of the parties involved; (c) the strategies of the actors, which aim for joint valorization at all stages of the chain of production and ensure the articulation of activities. The stages of integration concern both the pathway from the raw material to the production and trade of products and the successive valorization of scientific and technical knowledge necessary for the emergence and diffusion of complementary and cumulative innovations.

These notions are more globally addressed in the approach to economic and sustainable development and inclusion objectives set by international bodies (such as the sustainable development goals (SDGs) set by the UN: eradicating hunger and poverty, supporting sustainable agriculture and renewable energies, promoting sustainable development and education, conscientiously managing and preserving natural resources and ecosystems, combating social inequalities, etc.).

While the structure of innovation ecosystems in developing economies is very heterogeneous depending on the contexts visited, it is clear that they are often depicted as disorganized, fragmented or disconnected in their development process. However, according to the studies presented in this book, these systems are carried by actors of all categories who allow them to emerge, maintain and perpetuate themselves. The actions of actors, more or less formal, more or less interconnected, constitute significant dynamics that structure and shape innovation capacities. Represented by different forms of organization of collective action (producer groups, cooperatives, associations) and non-governmental organizations (NGOs), they are called upon to play a key role in the definition of policies and actions in favor of innovation and thus largely contribute to the formation and implementation of innovation capacities. They are sometimes called upon to complement or replace public authorities when public innovation policies are lacking.

But other actors are also stakeholders in local learning dynamics: civil society, with village communities, trades, genders, cooperatives, through different forms of learning, on the job, built by indigenous societies and communities mobilizing their own institutions of social cohesion, constitutive of local knowledge and collective social values.

This raises the question of the appropriation of this knowledge and the endogenization of local knowledge, combining culture, identity, confidence and a thirst for modernity. This bottom-up approach to capabilities, sometimes identified in situations of frugal innovation or the "bottom of the pyramid", is essential in understanding the current dynamics of innovation and learning in southern economies. The challenge is to be able to create learning capacities that reconnect with the resources that generate this learning as much as with the macro-economic conditions that territorialize it, and the sociopolitical locks or levers that activate it and allow it to be mobilized in the productive and social sphere. We can see, thanks to the studies presented by the authors of this book, that in countries where process engineering is quantitatively and qualitatively important, learning mechanisms are rich and can be observed by the amount of energy spent by actors to solve problems, but also by the capacity of actors to master new knowledge. Innovation here is the result of a set of efforts and trajectories of use of local knowledge, and the capacity to integrate the knowledge conveyed by foreign direct investment, international academic and agro-industrial cooperation or repatriated skilled labor.

The following issues are addressed by the authors in the light of the diversity of the countries and experiences studied: How do we protect the environment through new knowledge? What transformations should be carried out with the aim of improving the living conditions of populations? How do civil society actors participate in the formation of innovation capacities and in their implementation? How do public institutions integrate them into their operations? How can innovation processes, whether formal or informal, generate learning and innovation capacities? Engineering has a particularly revealing, mobilizing and integrating function for these capacities. How do different professional and rural communities form and structure knowledge that is the basis for learning capacities? What are the different forms of learning that are localized and specific to the capacities of developing country economies?

The book is structured around 12 presentations of experiences in different developing countries (Africa, Asia, Central America), and around the above-mentioned issues.

The first chapter by J.-L. Hornick, entitled "Big Changes in Global Food Security and the Issue of Development: Challenges and Hopes", discusses the importance of capitalizing on new discoveries to ensure nutritional security for future generations. In the current state of knowledge, global awareness of the issues and the implementation of known technologies must be able to delay the advent of a brutal decoupling of food supply and demand to respond to the important societal changes around food security and safety. The question is: How do we move from the potential for innovation announced by current technologies to the real and profitable application of these technologies? The transition between the "potential" and the "real" favors the emergence of agro-industrial sectors.

Taking into account the Natural Environment and global warming imperatively requires the application of a qualitative model capable of responding to the challenges of food, nature protection and wealth sharing. In Chapter 2, by P. Bouvier-Patron, entitled "Agri-environmental Frugal Innovation and Sustainable Development", the formation of agri-food chains must mobilize technology in a rational way to produce useful and adapted systems and products. This is the case for frugal environmental innovation, which relies on innovative creativity, whose main sources are "tinkering" and improvisation, within local communities of practice, to adjust to the constraints of the place and the time in the best way possible. The author presents the possible evolution of the cultivation of two world cereals that could be virtuous and indicative of the new model to follow: sorghum and rice. The observation shows that value chains are created or completed through frugal innovations.

The third chapter by E.M.F.W. Sawadogo/Compaoré and N. Sawadogo, entitled "The National Innovation System as Applied to Agriculture: A Methodological Proposal for Diagnosis in Africa", shows the legitimacy of the national innovation system (NIS) as a tool for policy analysis and governance in Burkina Faso. In this country, the NIS has acquired strong legitimacy in policy and research circles for some time. By borrowing a sectoral focus on agriculture and contextualizing a number of tools provided by the model presented, the proposal made by the authors consists of developing a practical methodological guide for the diagnosis of agro-innovation systems in Africa.

The fourth chapter by B. Ndiaye and A. Bayompe Kabou, entitled "Adoption of Rice Technological Innovation on Technical Efficiency in Senegal", analyzes the impact of the adoption of new rice technologies on the technical efficiency of farmers in Senegal. According to the latter, fertilizer treatment levels have more or less technical efficiency on rice farmers. In the end, Senegalese rice farmers who use fertilizer, improved seed and motorized equipment simultaneously are more economically efficient than those who only use fertilizer and improved seed.

Using the concept of the sectoral innovation system (SSI), S. Mathé, E.J. Fofiri and L. Temple attempt to understand and organize public policy choices in the cocoa sector in Cameroon in the fifth chapter entitled "Characterization of the Sectoral Cocoa Innovation System in Cameroon". Since 2002, there have been many stimulus packages in this country to increase cocoa yields and production through the transfer of scientific and technical knowledge. Nevertheless, their analysis reveals dysfunctions in this SSI and the value chains that could result from it. The authors propose to identify potential levers for a sustainable and more inclusive innovation trajectory for the production studied.

The sixth chapter, written by Y. Boughzala and N. Ben Mahmoud, entitled "Valorization of the Date Industry in Tunisia by Combining "Modern" and "Traditional" Knowledge and Techniques: Difficulties, Successes and Prospects", analyzes the date industry in Tunisia by highlighting the structuring of this agricultural value chain through strategic, economic and social issues. The objective is to understand the different impacts of these challenges on the organizations inherent to this economically promising sector for the country's exports. From this perspective, the authors show a dual agricultural value chain with a lack of structuring between the links in the chain, despite international leadership in terms of exports in value. Some challenges need to be addressed upstream to create a more efficient global value chain around a more equitable and inclusive sector. The public authorities and the various stakeholders must also play a more proactive role.

The seventh chapter, "Technology, Innovation and Sustainability of the Soybean Chain: Study of the Cameroonian Cotton Front Facing Environmental Challenges", by E.J. Fofiri Nzossié, D.N. Nitcheu Wakponou and C. Bring shows the dynamics of soybean production driven by growing demand from the national agribusiness sector and the cross-border market with Nigeria. This dynamic offers opportunities for the construction of a science and technology (S&T) policy in the agribusiness sector, on the one hand, and for the improvement of farmers' incomes, on the other hand, but it also responds to environmental issues in the face of the accelerated degradation of the country's Natural Environment. The authors analyze the environmental issues induced by soybean cultivation and show the accelerated regression of the vegetation cover and the degradation of the soil. The first issue is related to the expansion of cultivated areas through land clearing. The second is inherent to the intensification of the use of phytosanitary products and mainly glyphosate. Biochemistry would be a solution adapted to this type of problem.

In another sub-Saharan African country, the eighth chapter by B.W. Basse, S. Mbaye and O. Diop, "Impact of Good Agricultural Practices on Cashew Nut Crop Yields in Senegal", analyzes the impact of good agricultural practices in cashew nut sector support projects. They found that changing producer behavior is a good tool for increasing cashew productivity in Senegal. Thus, for a sustainable cashew sector, awareness-raising strategies can be developed by mobilizing a range of local actors (community radios, village chiefs, development agents, supervisory and research structures, etc.).

Good practices in the preservation of natural resources (especially plants and forests) are illustrated by H.L.T. Ranarijaona, T. Andrianasetra, L.J. Raharinaivo, V. Ramahatafandry, M. Befinoana, A.B. Ramiandrisoa, C. Maharombaka, S. Tomboanona, C.C. Totondrabesa, F. Andriamanantena, S.G. Andrianasetra, A. Andriamanantena and A.Z. Rabesa in the ninth chapter (Bioeconomy and Sustainable Conservation of Plants and Forests in Madagascar). The case of Madagascar is revealing in this regard: the creation of an innovative botanical garden with the involvement of researchers, engineers and the local population has formed an innovative ecosystem whose objective is to demonstrate the dependence of humans on traditional medicinal plants, and the new uses of plants, today in the fields of health and food, to strengthen conservation, the local economy and sustainable development.

The sustainability of innovation in agriculture can also be analyzed with reference to the concept of "bricolage". In Chapter 10, "Bricolage in the Agriculture Sector: Emergence Dynamic and Consequences in Vietnam", S.T.K. Le, T.T. Nguyen and P.A.T. Nguyen note that in most developing countries agricultural resources and means are scarce or insufficient, and that farmers in these countries, who produce mainly on a small scale, lack financial, skilled human and technological resources, and very often resort to simple and traditional, sometimes even unconventional, methods for cultivation or breeding. In this case, "bricolage" appears as a concept describing how individuals improvise by recombining existing, but individually less useful, resources to create value through creative and systemic reconstruction. The authors study the case of Vietnam to show that "bricolage" is a source of multiple innovations through the mobilization of relevant traditional knowledge and experience.

This "bricolage" is found in Sub-Saharan Africa, but this time in the centralization and marketing of agri-food products via "food hubs", which are concentrates of technological and organizational innovations. In Chapter 11, "The Contribution of Food Hubs in the Digital Age to the Sustainable 'Agri-food' Transition: A Review of Research in Sub-Saharan Africa", G. de la Paix Bayiha explains how to manage the growing gap between rural supply and urban demand for quality food products. This calls for a strong integration of agri-food chains to form sustainable "agri-food" systems at the territorial level. For direct sales to consumers, food hubs provide solutions for small farmers in Africa involved in sustainable agri-food models. These farmers can thus have easier access to urban markets and consequently improve the standard of living in the countryside.

Through a precise engineering approach, Chapter 12, by D. Andrianjafy, H. Rim Farasoa and F. Rasoarahona, entitled "Total Processing of Soy *Glycine max* through Valorization of the Tofu Whey in Cosmetic Products", deals with the study of the natural cosmetic sector in Madasgacar. According to the authors, the overexploitation of plant species threatens their existence, presents a potential danger for the ecology and, in the long term, will always have negative consequences for these countries. The valorization of cultivable plants is therefore a solution of choice. According to them, new cosmetic care products have been developed and this valorization enriches the value chain up to the production and marketing of soy cheese.

All in all, the contributions in this book offer a precise and in-depth reading of the mechanisms of agro-innovations for economic development purposes in the context of increasing food needs and accelerating climate change. New (or improved) socioeconomic models are emerging within specific organizations or sectors that value the importance of value chains created from localized learning processes, driven by do-it-yourself policies and constrained by developing country contexts that remain heterogeneous.

I.5. References

Asayehegn, K., Iglesias, A., Triomphe, B., Pédelahore, P., Temple, L. (2017). The role of systems of innovation in adapting to climate change: The case of the Kenyan coffee and dairy sectors. *Journal of Innovation Economics & Management*, 24(3), 127–149.

Casadella, V. and Uzunidis, D. (2017). National innovation systems of the south, innovation and economic development policies: A multidimensional approach. *Journal of Innovation Economics & Management*, 23(2), 137–157.

Casadella, V., Liu, Z., Uzunidis, D. (2015). *Innovation Capabilities and Economic Development in Open Economies*. ISTE Ltd, London, and John Wiley & Sons, New York.

Christopoulos, M. and Ouzounidou, G. (2021). Climate change effects on the perceived and nutritional quality of fruit and vegetables. *Journal of Innovation Economics & Management*, 34(1), 79–99.

Debref, R., Andreas Pyka, A., Perguiseppe Morone, P. (2022). For an institutionalist approach to the bioeconomy: Innovation, green growth and the rise of new development models. *Journal of Innovation Economics & Management*, 38(2), 1–9.

FAO (2022). *The State of Food Security and Nutrition in the World 2022*. FAO, Rome.

Gaitis, F. and Ouzounidou, G. (2017). Food safety: Strengthening the present with an eye to the future. *Journal of Innovation Economics & Management*, 24(3), 177–189.

Ortiz-Bobea, A., Ault, T.R., Carrillo, C.M., Chambers, R.G., Lobell, D.B. (2021). Anthropogenic climate change has slowed global agricultural productivity growth. *Nature Climate Change*, 11, 306–312.

Schlaile, M.P., Kask, J., Brewer, J., Bogner, K., Urmetzer, S., Annick De Witt, A. (2022). Proposing a cultural evolutionary perspective for dedicated innovation systems: Bioeconomy transitions and beyond. *Journal of Innovation Economics & Management*, 38(2), 93–118.

Touzard, J.-M. and Boutillier, S. (2017). Innovations and solutions for climate change. *Journal of Innovation Economics & Management*, 24(3), 3–8.

Uzunidis, D. (2019). A focus on food international challenges global food challenges related to climate change. *Agro-Innovation, Food Quality and Safety*, 10(1), 1–7.

1

Big Changes in Global Food Security and the Issue of Development: Challenges and Hopes

1.1. Introduction

Food security and safety are not recent issues on a human scale (Litzenburger 2016; Birlouez 2019). They have been the subject of relentless efforts since humans first became aware of the time scale in which life occurs. Control of these issues has developed in parallel with scientific and technological knowledge. A tipping point can be located around the middle of the 19th century, with the appearance of the first combustion engines that led to the increase in agricultural productivity and the transportation of products, and with the Pastorian discoveries that laid the scientific foundations of food preservation.

Five pillars of food security – in the broadest sense – have been chronologically accepted and formalized over the last 60 years: the availability of food in sufficient quantity, the possibility of economic access to resources, the conformity of sanitary quality and composition of ingredients and the plate, the stability of supply in the short and medium term and more recently, the long-term sustainability of the system to guarantee intergenerational solidarity (Akram-Lodhi 2009). These dimensions are of major importance to contribute to the stability of our societies. This chapter reviews some of the issues related to food security and safety, and avenues to ensure their continuity. More specific aspects related to food security will be highlighted.

Chapter written by Jean-Luc HORNICK.

1.2. Food security issues

Paradoxically, the major world conflicts were catalysts for population growth (see Figure 1.1). It was particularly after World War II that the human population showed a strong relative expansion. The phenomenon was most pronounced in Asia and later in Africa. This continent is currently the only one to present an exponential increase in population, compensating for the decreases observed elsewhere. World population growth has been linear for the last 50 years at a rate of nearly 70 million people per year.

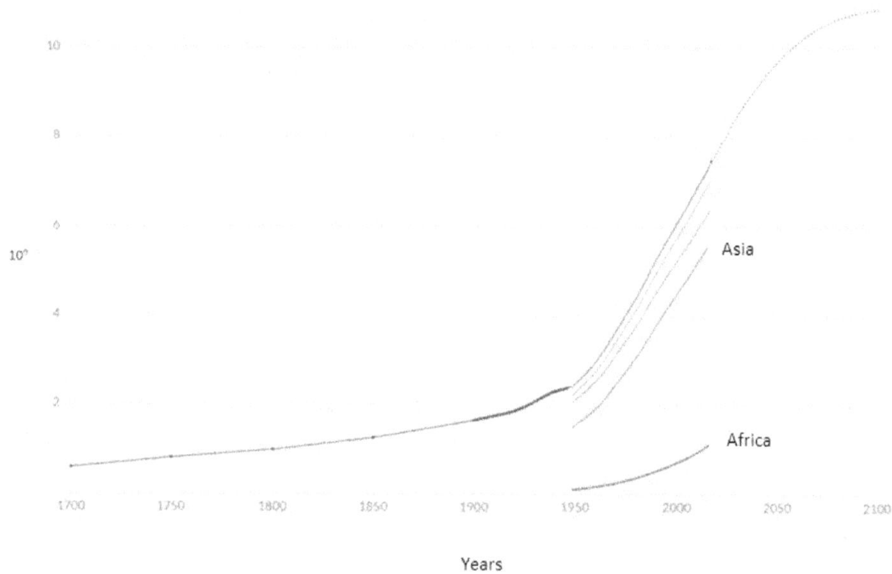

Figure 1.1. *Three-century evolution of the world's population, broken down cumulatively by continent from 1950 to the present day, and projected to the end of the century (Our World in Data 2021; United Nations 2019). For a color version of this figure, see www.iste.co.uk/casadella/agri-innovations.zip*

Improved agri-food practices and increased agricultural production have undoubtedly contributed to these demographic shifts. For example, the share of the world's population fed by agricultural practices using synthetic fertilizers has risen from virtually nothing to nearly 50% in 60 years (Erisman et al. 2008).

The preservation of food security is currently under threat, and humanity is facing challenges, which Hajkowicz (2015) has termed “mega-trends”, or large-scale conjunctural “shifts”. These shifts are all anthropogenic in origin and are likely to force humanity to reconsider how it will need to feed itself in future decades. The “demographic tide” movement mentioned above is one of these challenges.

The gradual increase in the earth’s temperature is another, as it threatens the very foundations of agricultural productivity. The National Centers for Environmental Information (NCEI) of NOAA (National Oceanic and Atmospheric Administration of the Department of Commerce in the United States) publishes a map of thermal anomalies recorded on the planet every year. In the last three years (see Figure 1.2), most regions of the globe have shown anomalous thermal averages compared to recent decades. Global warming will have consequences on the capacity of human organizations to ensure stable agricultural production. Thus, in most regions of the world, we must expect the emergence of extreme climatic events, such as torrential rains or scorching temperatures, which are not easily predictable in the short term. The management of water, whether to store it or obtain it, will therefore become a major issue of food security.

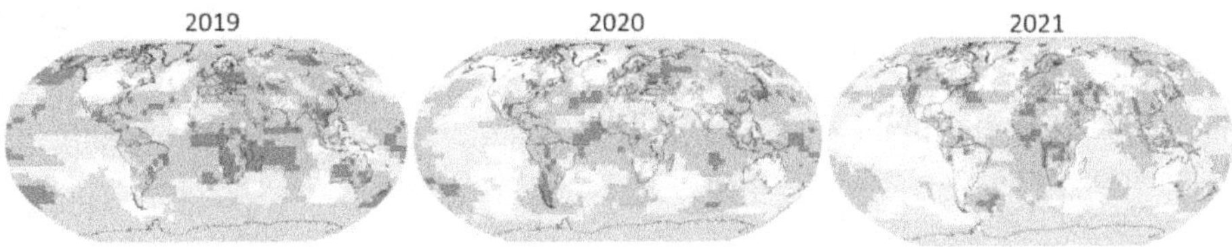

Figure 1.2. *Percentiles of thermal anomalies (positive in red, negative in blue) recorded over land in the last three years (National Center for Environmental Information, 2019, 2020, 2021). For a color version of this figure, see www.iste.co.uk/casadella/agri-innovations.zip*

Urbanization and migration to cities reflect another trend that has been sustained for several decades. It profoundly modifies the relationship between humans and the land, and causes an evolution from global, climatic food insecurity to a more economic form. Indeed, urban populations, although maintaining close social relations with peri-urban or rural areas, are more dependent on the availability of and access to imported food. Notwithstanding their density, they cannot reasonably rely on significant production of resources from the cities themselves. The vast majority of urban land is not suitable for agriculture due to the presence of roads and buildings. Even attempts to develop green roofs are hampered by the existence of artificial soil that is not deep enough to regulate water and nutrient cycling, as is the case in natural soils. As a general rule, the relative yields of such agriculture can only be low. Urban populations are thus dependent on rural production by road for their food, but even more so on imports through international trade routes, which are often facilitated by the presence of large cities near port areas, whether by river or sea.

In urban areas, privileged access to imported food resources has several consequences. On the one hand, people are no longer forced to cultivate the land, and, on the other hand, they have easier access to highly refined foods such as oils, simple carbohydrates or intensively produced meats. The latter are themselves often richer in fat – and therefore less rich in animal protein – because of the speed with which the animals grow. The consequences of access to these resources result in an increasing incidence of obesity when the income linked to the social class to which the individual belongs allows them to make these choices. Malnutrition through overeating is another major current trend.

Moreover, the easier access to imported food reflects another phenomenon that has become more widespread since World War II, namely, the speed and fluidity with which food travels along distribution channels. These two characteristics reflect a unified world in terms of transportation and communication, and facilitate the mechanisms of genetic introgression within ecological niches, and consequently, the transfer of pathogens, with the associated epidemic risks.

Paradoxically, despite these threats, the average life expectancy of the population is continually increasing, and although the natural age pyramid is normally presented in pyramidal form, as its name indicates, it shows a strong relative erosion of the proportion of the young population (see Figure 1.3). In some countries, such as Japan, there has also been a marked reduction in the birth rate (Matsuda 2020). Some pyramids may therefore evolve into the shape of an inverted teardrop, evoking the image of a hot air balloon rising toward a glass ceiling. Even if some states try to fight against the decrease in birth rates by promoting immigration or adopting family protection reforms, the global evolution of the age pyramid will have consequences on food security criteria. As people age, and particularly as they pass a threshold in older age, nutritional needs change (Rusu et al. 2020). Energy requirements – mainly related to the consumption of fats and carbohydrates – decrease, while the need for high biological value proteins, which can help kidney function, increases, as well as the need for minerals and vitamins, and the importance of fiber and antioxidant factors. The modification of the age pyramid profile thus gradually changes the composition of the plate. This is perhaps one of the elements that explains the current interest in organic or biological production.

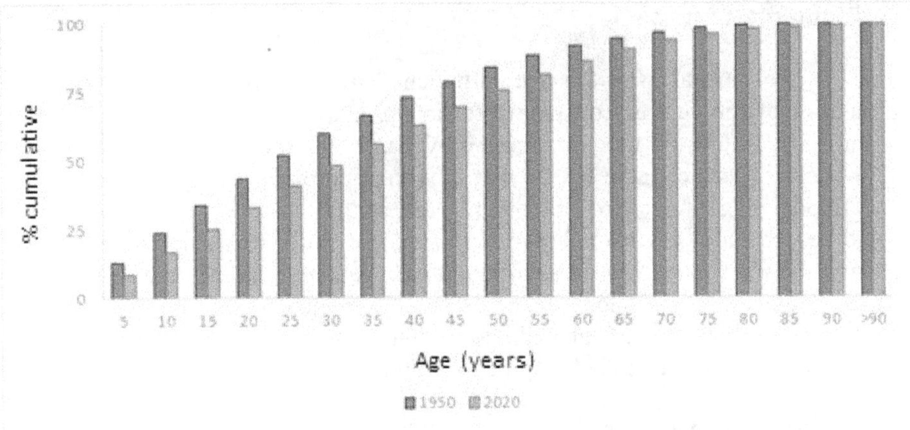

Figure 1.3. *Comparison of cumulative percentage of the world's population by age category, 1950–2020 (calculated from United Nations (2019)). For a color version of this figure, see www.iste.co.uk/casadella/agri-innovations.zip*

Taken together, these societal changes go hand in hand with a temporal contraction of food supply processes, but also with a form of spatial, economic and cognitive dilation of the steps in these processes. Indeed, with the development of means of transport, food travels further, which requires a multiplication of economic intermediaries and atomizes the distribution of added values – whether aggregated or not within these intermediaries – and this dilation also results in a feeling of insecurity as to the origin of the food. The latter is responsible for anxiety, and even chaos, when the already fragile confidence of the consumer in what they eat is called into question (Bricas 2003). Examples such as the mad cow disease or the recurrent discovery of undesirable residues in consumer products remain in the collective memory or are part of people's daily lives. This anxiety is rarely totally justified when compared to objective facts, but it is part of our human nature. However, the points of attention that mobilize societies vary according to culture. Thus, the countries of the European Union are particularly demanding regarding traceability and transparency within the food chain. Latin American countries, especially Peru, are more sensitive to issues of sustainability and the absence of xenobiotic residues in food products. The Asian region, on the other hand, is primarily concerned with controlling the food supply and biosafety, particularly through technological developments (Alltech 2020).

In light of these challenges, there is a case for looking ahead to how to ensure food security in future decades. The questions are not only about availability, but also about the other dimensions of food security and safety.

1.3. Elements of hope

There are several ways to meet the challenge of ensuring sufficient food availability in the future: consume less and produce more, or at least maintain the current capacity of food production systems. A human needs approximately 2,500 kcal of metabolizable energy per day and approximately 1 g of protein per kilo of body weight, or approximately 12% of energy intake. Half of these proteins should be of high biological value, rich in essential amino acids, and with a profile similar to that of human tissue. Proteins that meet these criteria are generally of animal origin. In a growing number of countries, humans can be considered to consume twice this amount for hedonic reasons. These are also partly responsible for the obesity epidemic in the human population.

To a first approximation, the animal and its food products are the result of a process of concentration of the proteins – often of plant origin – that it ingests, associated with a refinement of their quality. Ruminants are an exception to this rule, as the flora of their pre-stomachs are able to synthesize microbial proteins of very high biological value from proteins of mediocre quality, for example, from cereals, or even from non-protein nitrogen such as urea. However, the basic principle behind the construction of the individual remains similar: the animal extracts the essential amino acids that it cannot synthesize itself from food and stores them in its body proteins or exports them in products such as milk or eggs. The efficiency of this storage process is unfortunately low. In the best of cases, it is approximately 50%, for example, in genetically selected poultry. It is zero for non-productive animals whose feed consumption only allows them to maintain a stable weight. In addition to quantitative conversion, it is also necessary to consider qualitative issues. Indeed, the consumption of a cereal by an animal gives it a competitive character with respect to humans. On the other hand, coarse plant products are not, in the current state of our knowledge, a source of nutrients for the human species, but they can be perfectly valued by a ruminant to ensure, at a minimum, its survival. In summary, for every calorie ingested by humans, the calorie from animal products requires twice as much agricultural land as its plant counterpart. The reduction in demand for agricultural land can therefore result from a reduction in the consumption of animal products in favor of other ingredients. It is then based on a modification of the consumer's plate.

On the other hand, a decrease in food demand may be the result of better control of food losses and waste. Losses are often related to the initial and intermediate stages of production, for example in the case of crop diseases or animal epidemics requiring slaughtering. Waste is more likely to be related to losses at the end of the production chain, such as unsold products that are destroyed or food that has passed its sell-by date. Losses are more frequent in countries with low technological development, while wastage is more frequent in countries with food glut situations.

Waste can also occur at the beginning of the production chain in low-income countries. For example, slash-and-burn agriculture consists of cleaning and fertilizing soils with minerals by burning natural or crop straws. In addition to mortgaging the organic matter load of soils, this practice potentially deprives animals such as ruminants of a source of food, or farmers and herders of another form of resource utilization, if they had the technology to use it.

A second way to support food security would be to maintain and increase agricultural production and capacity. This approach seems conceivable only by increasing crop area or food production per unit area. This is a considerable challenge, and the second option is probably the most promising, both for environmental reasons related to the conservation of areas dedicated to biodiversity and for reasons of physical constraints.

The potential for increasing agricultural productivity is high. There is a logarithmic relationship between the application of crop protection agents, mainly fertilizers and pesticides, and agricultural yields. The latter can be increased 10-fold by moving from the worst cultivation conditions – lack of fertilizer, low water availability, low technological development and low farmer training – to the best. Belgium, for example, has a grain productivity of nearly 10 tons of dry matter per hectare, compared to less than 2 tons for most less developed countries (see Figure 1.4). The application of inputs, however, has adverse environmental consequences. Technologies to provide plants with essential nutrients in a "targeted, timely" manner to maximize the efficiency of their use, without losses through runoff or leaching, should be enhanced. In this regard, it should be noted that the African continent, with the possible exception of the Congo Basin, is largely covered with areas that are not very suitable for intensive agriculture, but are well adapted to extensive livestock farming. In other words, the continent can easily produce beef, sheep or goat meat, but at a slow pace, due to the low energy value of natural pastures. This should be taken into account in agricultural development policies.

Promoting technologies and their synergy and improvement can play a decisive role in achieving future global food security. Engineers have developed tillage techniques that are more respectful of soil structure. They reduce erosion and promote carbon storage in the soil. On the other hand, legumes can be valuable allies in enriching soil with nitrogen without the need for nitrogen fertilizers, due to their root structures that host rhizobial saprophytic organisms capable of transforming atmospheric nitrogen into plant-available nitrogen. Satellite technologies, on the other hand, provide valuable information to optimize land use in time and space. However, one of the greatest advances will probably come from revolutions in the genetic sciences. It is now possible to rewrite short excerpts of an organism's genetic code to try to improve its performance thanks to CRISPR-Cas9 technology.

On the other hand, gene editing is now reaching unprecedented levels of speed and execution.

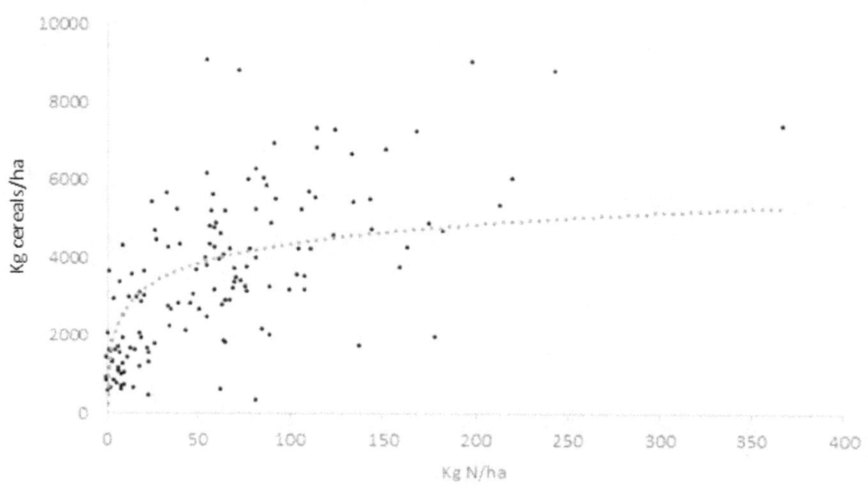

Figure 1.4. *National cereal production in the world as a function of nitrogen application per hectare (from FAO (2020)). For a color version of this figure, see www.iste.co.uk/casadella/agri-innovations.zip*

The first approach – genome rewriting – is likely to face many obstacles due to critiques – rightly or wrongly – of the use of genetically modified organisms. However, the technology is worth considering for the benefits it could bring (Rao and Wang 2021). For example, the photosynthetic efficiency of plants – currently close to 1% – could be improved. However, it would certainly raise the question of the risk of losing control over the propagation of plants that have become more competitive with their natural relatives. This simple example shows the caution with which these technologies must be used. Genomic editing, on the other hand, should make it possible to rapidly identify individuals – animals or plants – with a competitive advantage over their congeners. As founders of new lines, they would be endowed with resistance to biological or abiotic factors that are harmful to productivity. We can think of resistance to drought, high temperatures or critical diseases. However, such individuals could also generate lines that farmers or breeders would not have more control over in the case of genetic escape.

In this context, training, education and the transmission of knowledge will play an essential role. How can the fertility of the land be preserved? How can the ecological conditions of breeding be improved to protect animals from stress? How can food supply for animals be ensured with resources that are not competitive for humans? Some of the answers to these questions potentially exist in the endogenous knowledge of farmers, but they are not necessarily known and put into practice by farmers who are deprived of quality knowledge transmission. This endogenous knowledge risks being forgotten in the event of a generation gap following the migration of young people to cities and their progressive disinterest in agriculture and livestock breeding.

1.4. Conclusion

Food security and safety are strategic issues for peace. Their contribution to peace can only be guaranteed if the supply of food and the demand for it are kept in balance. Many means are currently available to us to reach this objective. Some of them are known and theoretically within reach. Seeking to optimize the consumption of animal products, for example, is concretely achievable on an individual basis, or even on a collective basis if food policies lead to it through awareness programs. The latter can also target the issue of waste. Controlling food losses is already more complex, as it requires better control of animal or plant diseases, or better access to modern agricultural technologies without increasing the pressure, in extenso, on the ecological environment. Scientific innovations, on the other hand, have been regularly modifying the existence of humans for more than a century and accompanying demographic growth. Their major importance cannot be denied, considering the proportion of the population that benefits from them. Nevertheless, we cannot deny the impact that they have had and continue to have on the planet due to the fossil fuel consumption that most of them require and the residues they generate. A trivial example is the production of urea for agricultural purposes. However, technologies under development or to be developed in the future suggest opportunities to improve global food security with highly targeted tools whose environmental impact could be negligible. It is not possible to make even a partially exhaustive inventory of these technologies in the context of this chapter. However, they could include, for example, viral vaccines using voluntarily mutated variants in a controlled manner, preventing the reversion of the agent into its pathogenic form, or products derived from messenger RNA technologies, which offer the prospect of unprecedented applications. In principle, these technologies require little or no energy. They rely on the capacity of the individual – animal or plant – to mobilize its own resources. They will therefore play a key role in reducing losses in the food chain and even in improving productivity. However, the question arises as to the extent to which these technologies will be under control, accessible to the greatest number of people, and not the object of a competitive trade that will leave out the poorest people or countries, as is currently the case in the context of the prevention of human coronavirus-related diseases.

Another issue, more related to food safety, concerns the problem of plastic packaging, whose impact on the food chain is disastrous. Unfortunately, they play an important role as microcontainers and guarantors of a certain hygiene, which makes them tools of social justice and accessibility to healthier food, especially in cities. Here, again, we hope that researchers will quickly develop technologies that are either able to replace these compounds in an ecological way or degrade them. This packaging is a reflection of our dependence on so-called "fossil" resources and on organic chemistry, the chemistry of synthesis, which is the mainstay of technological developments. This science is also an instrument of justice by allowing the transport and preservation of foodstuffs between producers and consumers who are far apart. Certainly, some countries have the opportunity to develop local, almost decarbonized agriculture, which could significantly contribute to covering the needs of their population for quality food. Others, on the other hand, would undoubtedly experience famine if they were to go down this path due to imbalances between population density and land productivity.

The dark side of this chemistry is having a growing impact on the climate and the environment, and it is therefore a priority to free ourselves from it as quickly as possible by developing ecological alternatives that are at least equivalent in economic terms. In this respect, it is difficult to see which sector other than research would be able to meet this challenge. Significant public funding would probably help to ensure a more equitable distribution of the added value generated by these future technologies.

1.5. References

Akram-Lodhi, A.H. (2009). World food security: A history since 1945. *Canadian Journal of Development Studies/Revue canadienne d'études du développement*, 3–4, 605–608.

Alltech (2020). Global Feed Survey [Online]. Available at: https://www.alltech.com/sites/default/files/GFS_Brochure_2020.pdf [Accessed 10 October 2021].

Birlouez, E. (2019). L'évolution de la perception de la qualité alimentaire au cours des âges. *INRAE Productions Animales*, 1, 25–36.

Bricas, N., Dury, S., Figuié, M., Maire, B., Delpeuch, F. (2019). Sécurité alimentaire et urbanisation : enjeux pour l'agriculture intra et péri-urbaine. In *Approvisionnement vivrier des villes du Sud : les enjeux et questions d'une agriculture de proximité*, Académie d'Agriculture de France (ed.). AGRIS (FAO) Editions, Rome.

Erisman, J.W., Sutton, M.A., Galloway, J., Klimont, Z., Winiwarter, W. (2008). How a century of ammonia synthesis changed the world. *Nature Geoscience*, 10, 636–639.

FAO (2020). FAOSTAT Data [Online]. Available at: http://www.fao.org/faostat/en/?#data [Accessed 15 September 2020].

Hajkowicz, S. (2015). *Global Megatrends – Seven Patterns of Change Shaping Our Future.* CSIRO Editions, Canberra.

Litzenburger, L. (2016). La sécurité alimentaire et sanitaire à Metz à la fin du Moyen Age. *Histoire urbaine*, 47, 131–148.

Matsuda, S. (2020). Progress of low fertility in Japan and other Asian countries: A theoretical framework. In *Low Fertility in Advanced Asian Economies*, Matsuda, S. (ed.). Springer, Singapore.

National Center for Environmental Information (2019). Assessing the global climate in 2019 [Online]. Available at: https://www.ncei.noaa.gov/ [Accessed 27 September 2021].

National Center for Environmental Information (2020). Assessing the global climate in 2020 [Online]. Available at: https://www.ncei.noaa.gov/ [Accessed 27 September 2021].

National Center for Environmental Information (2021). Assessing the global climate in 2021 [Online]. Available at: https://www.ncei.noaa.gov/ [Accessed 27 September 2021].

Our World in Data (2021). World population growth [Online]. Available at: https://ourworldindata.org/world-population-growth [Accessed 02 October 2021].

Rao, M.J. and Wang, L. (2021). CRISPR/Cas9 technology for improving agronomic traits and future prospective in agriculture. *Planta*, 4, 1–16.

Rusu, A., Randriambelonoro, M., Perrin, C., Valk, C., Álvarez, B., Schwarze, A.K. (2020). Aspects influencing food intake and approaches towards personalising nutrition in the elderly. *Journal of Population Ageing*, 13, 1–18.

United Nations (2019). Department of Economic and Social Affairs. Population Dynamics. World Population Prospects 2019 [Online]. Available at: https://population.un.org/wpp/Download/Standard/Population/ [Accessed 02 October 2020].

2

Agri-environmental Frugal Innovation and Sustainable Development

2.1. Introduction

The differences between individuals, countries and regions are first established by the satisfaction, or not, of basic needs (including food). A simple qualitative tool, such as Maslow's famous pyramid (1943), is sufficient to contextualize the problem. For OECD countries, consumption is often understood, beyond basic goods, as a quest to buy manufactured goods, mainly from the workshop of the world based on the low cost model that China has become, due to the strategic choices of private companies. The question of consumption, for the population of developing countries (non-members of the OECD), aspiring nevertheless for the same manufactured goods, is posed more centrally on the very "basic" satisfaction of fundamental needs linked in particular to food.

Certainly, the affluent fraction of a population in a poor country can be in tune with the affluent class in rich countries. On the other hand, a growing fraction of the poor in rich countries are losing access to basic needs through precariousness and are increasingly dependent on charity. Moreover, predictable but unlikely events (such as war or pandemics) in resource-rich territories, given the interdependence of economies and the intertwining of exchanges, only increase precariousness. It is therefore necessary to grasp things in their dynamic complexity, and therefore necessarily in a systemic way, by measuring the interactions and feedback likely to modify a fragile system, and this by means of small levers of initial incremental effects, in order to destabilize it and even create chaotic unpredictability or, on the contrary, to generate a more virtuous circle.

Chapter written by Paul BOUVIER-PATRON.

By considering the question of the "base of the pyramid" (BOP; Prahalad 2005), we are brought back to the essential question of how to satisfy the needs of the vast majority of the world's population, particularly with regard to food.

In particular, this means reconsidering a vision that has been hegemonic since the publication of The Wealth of Nations by A. Smith in 1776, based on the International Division of Labor, where the absolute advantage of each country would be to specialize according to the wealth of its initial factor endowments. Smith also admitted that the market is an institution and therefore a social construct (which has been perfectly established by D. North). The market is not the ether that bathes the economy, where the invisible hand corresponds to an automatic mechanism of spontaneous regulation. For purely ideological reasons, and in line with the interests of the United Kingdom in the 19th century, Ricardo, interpreting Smith, focused his analysis on comparative costs, implying that it was in the interest of every country to specialize in comparison with others where its factor endowments were most favorable. The comparative shift invites each country to enter into free international trade by exchanging, for example, grain (a product that the country needs) for other goods that it does not necessarily need from the point of view of the majority of the population, and assuming that the basic needs of the population are covered first.

Specialization, for comparative advantage, is all the more profitable with the use of machines. The technological challenge is then, in this framework, the search for the productivity of the factors (of all of the factors), the effectiveness (the end justifying the means), then the efficiency (centered on the return on the initial investment), to finally arrive at the selling in the optics of maximizing the selling price while minimizing the cost price by using marketing tools. Technical progress and therefore technological evolution play a decisive role in the competitive advantage (of a company, a region, a country, a zone), at the risk of a race to the top boosted by marketing, the purpose of which is so far removed that it loses all sense of the practical usefulness of the products or services sold.

On the other hand, the praise of slowness and local tradition cannot implicitly be what must be rediscovered at all costs and as the only meaningful element: customs and traditions contain practices, and only good practices are to be encouraged.

The selection of good practices is measured against global and local shocks. The issue of global warming – attributable to the Anthropocene (linked to the Industrial Revolution, the International Division of Labor and globalization) – is a major challenge to be met in addition to that of respect for the Natural Environment. Technological innovation must be a support to help find constructive solutions, therefore not abandoning it to the injunctions of marketing.

In itself, "the base of the pyramid" leads to a frugal approach and invites a reflection on the satisfaction of the needs of all by valorizing waste to produce useful and desirable products. This approach cannot be presented as a credible alternative without certain precautions: it must be possible to produce differently without using resources intensively and without damaging the Natural Environment. This open and collaborative approach (within communities of practice; Brown and Duguid 1991), from the bottom up (grassroots innovation; Seyfang and Smith 2007), must therefore, in order to find the appropriate technological level (or satisfactory in the sense of H. Simon) and an ad hoc organization, give priority to the design phase of an innovation (and eco-design, while potentially drawing inspiration from nature; Benyus 1997), through co-creativity using bricolage and improvisation (Cunha 2014; Bouvier-Patron 2021a, 2021b, 2023), by integrating the consideration of the Natural Environment from the outset. This approach is relative to the concept of environmental frugal innovation (EFI) (Bouvier-Patron 2020, 2023).

If this approach can and should be applied to all products of all sectors, it can be considered in the field of human (and animal) food. Food (which is an integral part of basic human needs) is the most sensitive to climate change and global warming, the deterioration of the Natural Environment, wars, diseases, bad agricultural practices, but also to bad consumption habits (thus adding, to forced underconsumption, the problem of malnutrition due to the loss of reference points and the disruption of traditional models via the effects of imitation, relative prices of foodstuffs and subsidized imports). Consequently, the inappropriate use of technology is adding to the vicious circle of poverty, malnutrition and disease. Food is therefore a witness to the functioning of globalization at the crossroads of all the contradictions and/or contemporary economic problems.

The opposition between food crops (human food) and fodder crops (animal food) is common, but crop plants can be dual-purpose; they can also, however, have a main purpose. Livestock farming is less and less about using animals as tools, although the plow horse or ox have an enormous potential for sustainable agriculture: notably for sowing, including technical practices by mobilizing a sowing machine, but also consider, in a less tedious way, the use of fertilizer for each plant rather than spreading it; dedicated technical devices, with training, are accessible to any farmer.

Mechanization, an expensive and increasingly technological investment, is rarely adapted outside of intensive agriculture (everywhere and for everyone), and may even be unnecessary or counterproductive (especially as a substitute for animal power, which is beneficial to the soil wherever it is used).

Mechanization focuses too exclusively on the elimination of the animal force and the organization of livestock exclusively destined for the production, with the help of technology, of milk and/or meat (certainly with possibilities of transformation by the agro-alimentary industry).

In addition, the power of vegetarian trends in the OECD leads to the revision of the arbitration between the areas reserved for grazing or for crops and, within the crops, between the share to be reserved for fodder crops compared to food crops. The debate can be extended to another level with the areas to be reserved for food crops (animal or human) and those dedicated to energy production (with agro-fuels).

In all cases, production, storage and distribution play a key role in the choice of crops and consumption: intensive production for large volumes to be stored and/or distributed globally does not have the same impact as local production for local consumption (short circuit). The technological level required is not the same.

Finally, at another level, but one that can rationalize the debate, global warming comes as a master and superior constraint dictating the harvest calendar and the choice of crop orientations, methods and/or their purposes. The equation to be solved is complex, with simultaneously: a rise in temperatures, generalized and increasing risks of drought, climatic hazards (floods, violent winds, intense cold, repeated and/or late frosts, heat waves), scarcity of water (or overabundance in the case of floods), reduction of water tables and their difficult renewal, depletion of natural resources and the collapse of biodiversity, impoverishment and/or degradation of soils (see Box 2.1), soil and water pollution (see Box 2.2) (water pollution resulting mainly from soil pollution by runoff or overflow or by deep infiltration), the contrasting risk-benefit of chemical plant protection products (fertilizers, herbicides, insecticides, pesticides), consumption patterns (often imported and standardized), problems of temporary or lasting shortages, storage difficulties, energy prices, dependence on technology, income variations, etc. In this very constrained framework, how can we propose a solution compatible with sustainable development to satisfy the greatest number of people with a constantly growing world population?

An essential question is whether improved seeds (hybrids) for crops that are commercially popular and highly valued by the agri-food industry are in line with the local needs to be met, and are a priority in the objectives of food for all. Beyond this, the other question linked to a more sophisticated stage is that of the stakes involved in the patentability of living organisms. Technology, for food, is thus a double-edged sword capable of the best (e.g. allowing access to water and in particular to drinking water or depolluting soils), but also probably of the worst (depriving local farmers of their right to cultivate seeds "freely", especially traditional ones).

Climate change is likely to put things in their place by prioritizing the obvious, often already known, but neglected or forgotten by pure commercial vision.

Soil depletion and degradation are the consequence of the deterioration or loss of ecosystems, clearing (with deforestation) for the extension of crops (often monocultures), mechanization, weeding (consisting of scraping the soil to destroy undesirable plants; in contrast, hoeing, which consists of aerating the soil without necessarily weeding to facilitate the root respiration of cultivated plants) and the nature and intensification of chemical inputs. In the latter case (Roose 1985), "mineral fertilization leads to soil acidification and groundwater pollution, irrigation favors the salinization of arid zones, phytosanitary control can lead to pollution, and mechanization sooner or later leads to compaction, physical degradation of soils and erosion". Soil degradation is therefore directly linked to intensive agriculture and/or intensive livestock farming and/or overgrazing. The phenomena of runoff (after intense rainfall on compacted, stripped and impoverished soils: therefore non-resilient) and erosion (by wind or water) result from this and are amplified by relief and aridity.

Box 2.1. *Soil depletion and degradation*

Soil pollution results from the degree of concentration of toxic chemical substances linked to the use of synthetic phytosanitary products, the passage of heavy machinery, industrial waste and the increase in artificial surfaces. This pollution is then diffused into the Natural Environment via water, air or living organisms (bacteria, plants, fungi) and causes disturbances in ecosystems, but also has an impact on human and animal health.

Box 2.2. *Soil pollution*

The issue of water is a priority, as emphasized by the FAO (2020). Technical and technological assistance must focus on access to water that can be used for crops (drinking water is another vital necessity for humans), but also for soil repair and depollution.

The combination of organizational innovation and technological innovation, adapted to the situation, is called environmental frugal innovation. It is therefore quite clear that agro-ecology is moving in the right direction and involves practices such as the increased use of organic fertilizers (whereas conventional chemical (synthetic) fertilizers come from non-living mineral sources through an artificial elaboration process) and the prevention of soil nutrient depletion. The more organic matter a soil has, the more potentially fertile it is.

In the hierarchy of issues and in terms of the relevance of its orientation, technology is decisive, with respect to the Natural Environment, on the issues of access and treatment of water, depollution, the provision of locally adapted tools, the quality of the production–storage–distribution–value of crops, energy and daily life.

Thus, let us recall that while maize (Portères 1955) has developed everywhere by dethroning local crops (in Africa, in particular), is water-hungry, intensively uses pesticides and is very sensitive to climatic hazards, sorghum is resistant to climatic hazards, does not require pesticides and needs much less water.

In this way, sorghum, of African origin, is gradually becoming popular and cultivated, notably in the United States (the world's leading producer of forage sorghum) and France (with the objective of reducing the dependence of French farmers on transgenic Brazilian soybeans, which, in addition to its origin, leads to massive deforestation of the Amazon).

Forage sorghum is used as animal feed or as silage or pasture in western countries, but has other potentialities, such as energy production (agro-fuel or biomass).

Several types of sorghum are used in Africa (Cruz et al. 2019): forage sorghum (after silage or direct grazing by animals), grain sorghum (for human food, but also for animal feed) and sugar sorghum (valued as fodder or for sugar or alcohol production).

The choice of species and seed, by type of sorghum, depends on the available water resources and the type of agriculture (extensive or intensive) mobilizing different inputs (organic or synthetic). With farmer seeds in Africa, if the yield is modest, it is stable because it is well adapted to climate change (Beucher and Bazin 2012). Moreover, the yields of organic crops can be equivalent (or slightly lower depending on the study) and, in some circumstances, higher (see: https://rodaleinstitute.org).

After a presentation of the context and general issues of technology and food where we will question the satisfactory technological level, we will examine two crop developments related to food: sorghum in Africa and rice in Thailand.

2.2. From HighTech innovation to a satisfactory technological level

What we are currently observing has already been explained in the Meadows report published in 1972, when climate change was not even mentioned. There is therefore a magnifying glass effect, today, by a variable that was already at work, but previously hidden: global warming.

The growth of the world population and the lifestyle aligned with the OECD is taking place for a planet that remains the same size. The "base of the pyramid" questions an intensive model that uses too many resources and is based on the imitation effect. Thus, the question is less to deprive the African consumer of wheat imported from Ukraine or Russia than to submit other possibilities such as local products and show them the commensurability of the price of food goods: the short circuit of local products, giving income to the farmers and covering food needs does not have the same global cost and effect on the Natural Environment as imported and subsidized products. The access price, by reasoning in full costs, discriminates between what should be expensive and what should be cheaper. Local food products have more nutritional value and more health benefits than imported products.

The wealthy class that desires imported product for the sake of snobbery and imitation of the "Western Model" can and should pay a high price, especially as they do not play by the local model.

The question is, therefore, rather organizational in the sense that the issue is to know how local farmers are able to feed the cities (Basquin et al. 2014; Jacquemot 2022) where the population is now flocking. On the other hand, access to water and soil, energy, the quest for a level of production (respecting the Natural Environment without depleting resources), the management of climatic hazards, storage and distribution, and then marketing are as much organizational problems as technological ones. Solar energy is surely the most decentralizable locally (likely to dispense with network infrastructures – as long as they already exist and/or are functional), and by freeing itself from natural barriers or distances, would be crucial for production (domestic, peasant agricultural, artisanal and/or in small workshops or small businesses), as well as for everyday use (access to drinking water, extraction from wells, possible sanitation treatments, sanitary water, food cooking, lighting, cleanliness, sanitation, waste treatment, means of communication including telephone and Internet, short distance travel, motor energy for the most common tools, etc.).

In this sense, local initiatives, supported by NGOs and in relation to research centers or groups of the OECD, are the place of real exchanges and effective imports to be realized. Technology must make it possible, in the medium term, to do without fossil fuels or, in the short term at least, to reduce their use by improving the thermal or diesel engines of generators, light or heavy vehicles and those of the agricultural machines used (from a grain husker to crushers or mills for grains, to a small tractor or larger machines).

This is a primary orientation as long as the solar energy, which respects the Natural Environment, can be decentralizaed and appropriated locally, to supply batteries (equivalent in power to replace diesel engines and their autonomy), but will not be technically operational and economically viable for weak incomes (in comparison with the OECD purchasing power).

This said, the scope of innovation on technical systems in extreme conditions of use (dust, heat, humidity, high thermal contrasts) is particularly expected and cruelly lacking (which can be explained by the weakness or absence of perceived profitability for companies with R&D capacities, knowledge and know-how). It is therefore likely that the coverage of this technological need will be essentially the result of initiatives by "Makers" (Anderson 2012) within "communities of practice" (Brown and Duguid 1991) via (Gershenfeld 2005) "FabLabs" (supported or accompanied by companies for the industrial stage: either in a commercial approach or in a philanthropic approach, but good for the image and reputation). The "Maker" designates the creator-manufacturer for a reappropriation and return on the making and the know-how by mobilizing the technology in a creative and co-creative way to satisfy individual or collective needs locally. The "FabLab" (space of experience and realizations) is a place facilitating the exchange of knowledge and know-how, designing and realizing, with the help of computer and digital technology, all the conceivable useful and adapted objects.

Allowing decent living conditions, including affordable and healthy food, while respecting the Natural Environment from the perspective of sustainable development, appears to be sufficiently federative throughout the world for an effective and substantial impulse to exist. The orientation toward useful products that are robust, simple, locally repairable and affordable is an essential issue of the frugal approach.

The adapted equipment, more or less technological, as well as its maintenance, does not necessarily have to be purchased by each farmer themselves (with or without credit), but can be collectively owned within a cooperative, which allows for the sharing of charges and costs.

For agricultural crops, technology is a specific element, among other variables that can be thought of independently (see Table 2.1).

By grading the importance of each variable according to a given context, a profile is revealed that can, among other things, show the importance played by technology in farming activity. The technology used for cultivation is not intrinsically oriented, but is orientable in its HighTech versus LowTech configuration.

Then, there is no impediment to achieving resource-intensive, low-mechanized agriculture for export with little technology and no environmental concerns.

Versus	
Non-agri food	Agri food
Exportation	Local subsistance
High yield	Reduced yield
Intensive mobilization of resources	Sobriety
Non-organic cultivation	Organic cultivation
Use of phytosanitary products	No use of phytosanitary products
Use of synthetic phytosanitary products	Use of organic phytosanitary products
Low respect for the Natural Environment	High respect for the Natural Environment
Globalization-oriented public policy	Food self-sufficiency
High degree of mechanization (mechanical strength)	Weak mechanization (animal or human strength)
Storage capacity	No storage capacity
Long distribution channel	Short distribution channel
Industrial seed with proprietary patent	Local/exchanged traditional farming seeds
Dependence on fossil fuels	Low dependence on fossil fuels
HighTech technology	LowTech technology

Table 2.1. *Variables related to farm model choices (source: author)*

On the other hand, it is possible to carry out an agriculture respectful of the Natural Environment (ideally “Organic”) to satisfy exportation (as can be the case in Madagascar, including for aquaculture with shrimp with organic labels) and/or local needs (for a qualitative food self-sufficiency of territories articulating villages and cities) by falling within the framework of sustainable development.

Each time, the question may arise of mobilizing, or not, the most high-tech technology. It is as much a matter of financial resources, necessity and usefulness (as is the case in Israel, for example, in the Negev desert for the production of portable water, as well as systems for the irrigation of crops).

This neutrality of the technology shows the importance of its employment doctrine, the financial means available, the importance of communities of practice, the role of co-creativity and the knowledge and/or skills available. The issue is therefore the diffusion and decentralization of the technology at the local level. This dissemination/decentralization can reproduce a model of dependence on western technical advisors. However, in the absence of local appropriation of all aspects (from design to use), we can at least think of easy availability in terms of usage and access price, a possibility of use without conditionality of external aid and a capacity for local “reparability” in case of problems. Expertise, competence, know-how, bricolage and improvization (Cunha et al. 2014; Bouvier-Patron 2021a, 2021b, 2023) must be encouraged and stimulated at the most decentralized level.

The field of investigation concerning technology that can be appropriated by third parties who need it (e.g. solar energy and its derivatives) is part of a mindset specific to "Makers" (Anderson 2012), in communities of practice (Brown and Duguid 1991) and within FabLabs (Gershenfeld 2005), as a node of partnership and transmission of a technology that is accessible to all, rather than an imported model with a purely commercial orientation centered on the challenge of export markets.

This orientation of research on technologies that support agricultural activity also and above all encourages us to ask ourselves the question of the satisfactory technological level. The need, in developing countries, is not for turnkey HighTech solutions that cannot be appropriated, as would be the case with products protected by proprietary patents, leading to dependence and confinement in a technological straitjacket with uncontrollable effects.

Substituting local seeds or types of crops with imported ones is in itself a problem (Africa's dependence on European wheat is significant of a drift), but increasing dependence by injunctions linked to imported technological choices as for genetically manipulated seeds (hence the term: genetically modified organism (GMO)), whether for the local market or for export, is an additional step toward a leaching of possibilities by a system of imprisonment of use with effects of irreversibility (on the Natural Environment and biodiversity), as well as compulsory practices (those of the use of synthetic pesticides dedicated to the seed) to the detriment of all the knowledge and adapted know-how.

The link with the climate crisis is real when we cross the perspective of an economic model with deforestation (following clearing for wood energy, an intensive culture and the exploitation of tree species) concerning primary forests (green lungs of the planet). The cultivation of transgenic soya in Brazil (intended to supply animal feed in France) is directly linked to deforestation.

However, the disaster is not unique to GMOs, since the substitution of local production by export production such as palm oil in Indonesia leads to the same result to serve the world economy.

Indonesia is therefore a leader in the production of palm oil for the agro-food and agro-fuel markets, but at the cost of deforestation (at least a quarter of the country's deforestation is linked to palm oil harvesting).

More globally, the industrial logic of agriculture favors hybrid seeds selected for yield (and therefore productivity), but high-yielding varieties penalize the genetic richness of seeds, affect biodiversity and ecosystems, are energy intensive (especially fossil fuels), pollute soil and water, generate *carbon dioxide* (CO_2) emissions and require many synthetic phytosanitary products (fertilizers, herbicides and pesticides). The latter being dedicated to selected hybrid seeds, this forms an economic system of a proprietary solution where "everything" must be bought (which amounts, by design, to making the farmer dependent by trapping him first by the prospect of yields – probable in the short term, but delicate to achieve in the long term). This logic, driven from outside, disturbs or destroys local knowledge and know-how, and is irresponsible from the point of view of sustainable development. To reverse this situation, the role of technology would consist, around an organizational innovation that selects good practices, of (a) developing non-chemical inputs (fertilizers, herbicides, insecticides and pesticides) of the biological type possibly produced by the farmers themselves (but more likely by local companies (such as Bio Phyto Collines, see: biophyto-benin.com), which have the expertise to propose effective products that are no more expensive than synthetic products) in regard to reinforcing or complementing local organic phytosanitary products; (b) proposing technical solutions for use (such as microdosing; Aune et al. 2007) or recommendations for use according to the place, period, situation or calendar.

From the perspective of serving food to all, the frugal approach is an invitation not only to reuse waste, but also to engage in a reflection on a clean technology that only generates by-products that can be reused indefinitely (Cradle to Cradle (CtoC); Braungart and McDonough 2002).

It is therefore necessary to select good practices for food for all that are compatible with sustainable development (whether they are traditional, such as crop rotation – always leaving an area fallow – with one main crop for commercial purposes and another for a secondary crop, but intended to cover the soil and enrich it, to repair it or for grazing; or modern, such as the microdosing of appropriate organic fertilizers). This approach implies long-term thinking and moves away from the usual objective of the short-term profitability of an economic activity.

Even under water stress, combining an adapted cereal crop (such as sorghum) with a vegetable allows us to reach a yield equivalent to that of a monoculture, but with an additional harvest of protein-rich legumes. This type of cereal/legume association optimizes the use of water resources due to the beneficial effect of the legume on the fixation of resources and a limitation of the evaporation of water from the soil due to a denser plant cover. Food security and maintenance of soil fertility are achieved without wasting resources (especially water).

It is therefore the design, very early on, of a clean technology that does not have an effect on the Natural Environment and leaves by-products that can be reused indefinitely, which presides over the EFI, necessarily containing a part of the organizational innovation necessary to obtain a product innovation respectful of the Natural Environment.

2.3. Sustainable food in a necessary context with respect to the Natural Environment and global warming: sorghum in Africa and rice in Asia

Self-produced farmers' seed varieties still constitute 70%–80% of those exchanged between African farmers (CIRAD 2013). These seeds, which have the widest genetic heritage (Clavel 2016), are therefore likely to have a plasticity that allows them to respond better to climatic challenges (from hazards to the very issue of global warming leading to heat waves, droughts and/or floods) and sustainable development by saving resources and minimizing inputs (especially the artificial ones proposed with industrial seeds).

However, industrial seeds (selected and listed for certification for sale), called hybrids, are more or less threatening to Africa.

In addition to industrial seed companies, the Gates Foundation (the Bill and Melinda Gates Foundation or BMGF has been in existence since 2000 and is now endowed with $70 billion as capital in a profitable investment fund, but in sometimes controversial sectors, to reinvest (only) dividends in finance, by philanthropy (or to control and direct or influence), global projects aimed at improving food, health, education, human living conditions) advocates for the globalization and intensification of agricultural crops. Philanthropy masks techno-science strategies for Africa as an extension of the prevailing OECD model. This foundation funds the African Green Revolution Alliance (AGRA) program, which, according to its experts, estimates that improved (or hybrid) sorghum seeds would allow for yields of 3–4 tons per hectare in West Africa (compared to 1–1.5 tons per hectare for farmers' varieties). However, the purchase of hybrid seed is already a very high price for smallholders (compared to the usual use of traditional seed). Moreover, the shift to intensive production almost necessarily implies the replacement of reproducible traditional seeds by the hybrid species that farmers must necessarily buy every year. The freedom of the farmer (this is already the case within the OECD) is reduced by this obligation to buy, and this is a strong tendency (independently of the derogations or exceptions) due to the reinforcement of the legislation under the pressure of the lobbies tending to favor intellectual property (quite debatable concerning living organisms since the sanitary argument – the guarantee of harmlessness offered, of the seed or the plant, by a certification, which is in itself a good thing, has a confiscatory counterpart of the peasant's freedom by making them a slave of the seed producer). To exchange or commercialize seeds (and their derivatives, such as seedlings), they must be registered in an official catalog of species and varieties as certified seeds (*Certificat d'Obtention Végétale* (COV), a specific form of intellectual property, distinct from patent law, according to the criteria of distinctness, homogeneity, stability (DHS)). Clavel (2016) insists on the paradigm shift operated in Europe after 1945 by "a private seed supply of genetically homogeneous and traceable varieties"; selected and homogeneous varieties to the detriment of the freedom to cultivate varied and diversified ancient seeds whose rich genetic heritage allows adaptation to places and situations.

Africa was temporarily sheltered as long as intensive agriculture, promoted everywhere (and misleadingly named Green Revolution in developing countries), was circumscribed in zoning and type of cultivation (then allowing the maintenance of local family agriculture); the large international organizations and big companies focused on large farms of export crops and considered hierarchically more important (wheat, corn, cotton in particular), while neglecting the indigenous crops that were considered secondary (indigenous food crops: sorghum, millet, cassava, yam, plantain banana, cowpea).

Little is known about cowpea, which is one of the main legumes produced in Africa, from the lentil family, but similar to European beans.

The shift to the intensive model in Africa is taking place within the framework of the WTO through the establishment of common standards and involving, in particular, adherence to a regulatory system (likely to deprive local farmers of the freedom to sow, exploit and market). The regulatory system that applies in Africa is none other than the European system UPOV (*Union pour la Protection des Obtention Végétales*: Union or the Protection of New Varieties of Plants): it consists, for each type of variety, of offering hybrid seeds in a catalog after certification. This system, in its most restrictive version dating from 1991, applies in Africa and prohibits local farmers from using self-produced seeds outside their own fields. Thus, registration in the catalog constitutes a high technical, financial and legal barrier to entry that is particularly insurmountable for any local farmer wishing to market a traditional seed.

Beyond the question of intellectual property (and the associated HighTech) on living organisms, the most important thing is to question the meaning of things. Is the goal to make profit thanks to living organisms or to answer the question of a healthy food for all that is respectful of the Natural Environment?

If the answer is the second possibility, the innovation effort, already mentioned, must focus on decentralized means locally to access resources without depleting them, but also to produce, store and distribute while respecting the Natural Environment.

To concretely understand the desirable evolutions allowing us to respect the Natural Environment and answer the challenge of global warming, let us take two examples: sorghum, a plant of African origin also consumed in Asia (and which is increasingly successful in Europe and in the United States for animal feed), and rice of Asian origin, but also consumed in Africa.

2.3.1. *Sorghum*

As all phenomena (economic, social, military or climatic) produce fluctuations, Africa's own exported commodities are also subject to world prices. This is true for food, but it is also true for logging, mining and industrial extraction (private or under public license).

In Africa, sorghum tends to be replaced by European wheat (or even Asian rice). It does not matter what the cause is at this point (the result of the greed of the firms coming from exporting powers, the doctrines of international organizations, the business of local elites or political discouragement) because the cause does not change the situation of Africa's dependence on the basic goods it needs. Everything that is not produced locally depends on international volumes, qualities and prices (whose springs are external to Africa), and this permanent subjection to external conditions, beyond dependence (always multiplied), weakens the populations.

In Africa, imported wheat (especially from Europe, where it is massively produced with the help of the common agricultural policy (CAP)), which is sometimes subsidized locally as well, is generally a cheap cereal to buy (and less expensive than local products). Wheat flour is reputed to be easier to work with. Wheat bread is now very much appreciated by the local urban population. It is therefore easy to import wheat, while traditional bread made from local, non-imported cereals is abandoned. However, African bakers could offer special breads with mixtures of flour (wheat, millet, sorghum, soya, cassava or others) in varying proportions: a range of varied products to satisfy all customers and at all prices.

The training of artisans, access to appropriate machinery, and food and nutrition education (or re-education) for healthy and balanced local food is a long but necessary learning process.

Cereals in Africa are diverse but, seen from the outside, are too often reduced to Mil.

To understand it, millet designates several Poaceae (grasses) whose seeds are small (small grain millet or pearl millet: Mil) or large (large grain millet: sorghum), cultivated especially in arid and dry regions (in Africa and Asia): it is a nutritious food, often consumed in the form of patties.

The trap of soil impoverishment (and therefore the impoverishment of the local population), which we must get out of, is that of intensive monoculture for export. If exploited soils are degraded (and emptied of their organic matter and their nutrients) in the arid zones of Africa or Asia, we have seen that restoration is possible by cultivating better through good practices without spending more, but earning more. Farmers must be able to valorize natural resources in a sustainable and autonomous way through good practices: organic management of soil fertility; crop-livestock association (livestock contributes to soil fertilization, and milk production in particular, and increases the range of valuation options for small farmers) and/or crop association of cereals and vegetables (crop rotation and intercropping with nitrogen-fixing legumes); plowing by animal traction rather than mechanically and by limiting its frequency; conservation of the biodiversity of flora and fauna and ecosystems.

These practices must be learned and disseminated, but also updated and exchanged within local communities of practice. Producing better while protecting the Natural Environment in the context of global warming is an appropriate way to maintain and enhance local production and secure land.

Sorghum's properties are interesting for its cultivation: for photosynthesis, some varieties have a cycle shorter than three months, which allows several harvests per year instead of one; for drought, its resistance is remarkable (which is a considerable asset from the point of view of global warming); its water requirements are lower compared to corn, and its cultivation also requires, comparatively, much less fertilizer (or even no synthetic fertilizer), as well as less work (for the same quantity of harvest).

Beyond the food issue, the Intergovernmental Panel on Climate Change (IPCC) encourages farmers in Africa to favor indigenous cereals such as sorghum, which are more resistant to drought and can adapt to increasingly frequent and brutal climatic hazards.

Under "ideal" conditions (i.e. disregarding the depletion of resources: scarcity of river or rain water; soil poverty; soil or water pollution; deforestation), yields are, on paper, higher for wheat and maize, based on intensive cultivation using hybrid seeds, consuming water and requiring chemical inputs. In real conditions and without artificial expedients, cultivation based on local sorghum farmer seeds, economical resources and cultural interventions while preserving the natural environment and supporting global warming, offers acceptable yields (and can be improved by good practices and good organization of crops).

The problem is therefore not so much the yield as much as the promotion of local products among local consumers who, because of ease, snobbery or obtuse imitation of a so-called diet (of the western type, but badly dosed and generally leading to malnutrition through caloric, carbohydrate or lipid excesses), turn away from a healthy food plate based on local products.

Hence, it is important to organize local farmers through local, regional and national associations and cooperatives, while also becoming involved in the local distribution and selling circuit.

From a nutritional point of view, sorghum has many advantages: it is naturally gluten-free, provides starch, protein and dietary fiber, and is rich in iron, calcium and phosphorus. Its glycemic index is moderate. Sorghum is also known locally as a preventive agent against various gastrointestinal pathologies.

In Africa (Fusiller 1994), it has been shown that the sorghum versus maize match is tilted in favor of the latter under conditions of ease of cultivation and artificialization, but there is also an incentive to substitute sorghum for maize to export cotton. Indeed, corn allows for a higher yield in a smaller area (which frees up space for cotton). However, this doctrine of quantitativism and internationalization based on the model of intensive cultivation and export (which has the probable effect of improving the country's ranking and attracting positive notes from international organizations better disposed to lend or finance) induces the destruction of local agriculture, ancestral practices (often beneficial at several levels), the Natural Environment, resources, biodiversity and ecosystems. It is thus still and always a question of sacrificing sustainable development on the hotel of the apparent short-term profitability. This increases the dependence and submission of the local economy to world prices and relies on hypothetically stable resource conditions without anticipating economic, social and financial crises, as well as the depletion of resources, the destruction of the Natural Environment and global warming.

On the other hand, Jacquemot (2021, 2022) advocates, from an optimistic and sustainable perspective for Africa, a model of territorial articulation between town and country that values agricultural modalities. It enables escaping vicious cycles while encouraging food progress to fight against hunger and malnutrition. The transition toward an agriculture and a breeding respectful of the Natural Environment, in Africa and elsewhere, is not an option but a solution to take up many challenges, such as satisfying local food needs while apprehending global warming.

The other vision, that of intensive cultivation with improved seeds or those resulting from GMOs and synthetic inputs, arguing that only quantity counts and that Africa holds the largest potential reserves of arable land, is a dangerous utopia whose price is the sacrifice of the Natural Environment, biodiversity, ecosystems and sustainable development: would Africa be a reserve of unused arable land after total clearing, destruction of natural reserves, deforestation and expulsion of flora and fauna?

2.3.2. *Thai rice*

Rice is a food consumed everywhere but produced centrally in Asia and particularly in Thailand. Food self-sufficiency seems to be achieved in Asia in general and in Thailand in particular, but the "way" of development is in question.

Rice is as much a food resource in Thailand as an ancestral practice, a way of life and a cultural value.

More broadly, the "Thailand 4.0" strategy (Jones and Pimdee 2017) aims to develop value chains and innovation so that the country moves out of its intermediate status to become more competitive and attractive, including HighTech. An innovation ecosystem on food is encouraged.

Concerned about its agricultural and rice heritage, Thailand has also adopted specific legislation for European-inspired protected geographical indications (PGIs) (Ministère de l'Agriculture et de l'Alimentation 2019).

The European Commission (EC) has even accepted the registration of an aromatic rice under PGI: *Hom Mali* from *Thung Kula Rong-Hai* (area in the north–east of the country, where the land is irrigated by the monsoon and the Mekong River), reputedly the best in the world. This Thai long grain rice has a fragrance of jasmine (*Mali* in Thai), linked to the soil (therefore not linked to the jasmine flower).

That said, Phélinas (2010) notes that Thailand has experienced the double shock of the arrival of new export competitors and the depletion of water resources, given that rice cultivation requires significant water needs. Reduced rainfall, waves of drought, failures in the water delivery system, urbanization and internal competition over water use are jeopardizing rice cultivation as the uncertainty over available water discourages investment (and, consequently, innovation).

Rice cultivation in Thailand has been able to maintain itself because it is a high-demand export product with a stable or rising average world market price, while the downward fluctuation of the national currency exchange rate (the baht) reinforces the competitive advantage in exports.

The lack of labor is more worrying: the rural exodus to urban sites and the transfer to industrial activities lead to less availability for agriculture and a higher cost of labor. The result is a lower profitability for exports compared to other countries where the agricultural population remains abundant and cheap.

Mechanization to replace labor and to irrigate or drain rice fields (depending on the needs of rice cultivation) is a theoretical solution that only becomes practical and effective with local control of the appropriate technology, the dissemination of knowledge and know-how, and affordable prices for machines (in addition to access to credit or rental); this equipment can also be made available within cooperatives (as is the case for maintenance). Credit poses other problems of repayment for incomes that may be uncertain; rental must be developed and could be the subject of a national policy.

Today, Thailand's apparent productivity is lower than that of other rice-producing countries. However, the lower productivity is explained by the fact that Thailand favors, by its history, the cultivation of more qualitative seeds.

In addition to this PGI rice, the opportunity for the revival of a more qualitative national rice production lies in the paradoxical fact that a large part of the country's rice production does not benefit from natural river irrigation, but relies on rainfall (and even water reserves when it is possible to store water). Indeed, high-yielding hybrid seeds, as well as chemical inputs related to these crops (fertilizers, fungicides, pesticides, insecticides and synthetic herbicides) present a greater economic risk as soon as water is scarce, which should make their use in most rice fields less attractive. In the short term, this makes it more difficult to operate, but in the longer term, it is a better guarantee against soil and water depletion, degradation or pollution (resulting in particular from the direct and indirect consequences of the use of phytosanitary chemical inputs). It remains that the seeds must be well adapted because the inadequacy that condemns them to attacks by diseases and pests results in the need for treatment by inputs (generally chemical).

A quality approach is a favorable commercial positioning in a logic of export on a niche market (compared to the mass of rice sold and consumed in the world), but also from a perspective of sustainable development, because what is less known is that the artificial model of intensive rice farming is not neutral from the point of view of global warming with high methane emissions (second only to livestock). Methane is a greenhouse gas with a more powerful warming power than CO_2, which is used as an international reference to analyze global warming.

The aim is to encourage producers to abandon the intensive model and all bad practices, and apply new ones that emit less greenhouse gases. Rice farmers are thus encouraged to use appropriate seed varieties and cultivation practices or techniques (soil leveling, alternate water management (by drying), application of organic fertilizers after soil analysis and use of rice straw and stubble to avoid burning).

The commitment made by Thailand is to reduce its greenhouse gas emissions and therefore grow crops that are more respectful of the Natural Environment. At this stage, rice farmers are locked in a vicious circle: affected by climate change (which causes droughts and floods), they themselves contribute to this climate disruption because, in addition to the non-respect of the Natural Environment, intensive rice fields release, at the same time (and in quantity), greenhouse gases.

The main solution (La Croix 2019) is not technological per se, but an organizational and practical innovation that is spreading throughout the world: the "System of Intensive Rice Cultivation" (*Système de Riziculture Intensive* (SRI)) developed in 1983 in Madagascar by Henri de Laufanié (1920–1995: French agronomist and Jesuit priest). Misnamed, the intensive rice cultivation system promises better yields, but this practice is the opposite of intensive cultivation. Therefore, to avoid ambiguity, we prefer the term innovative rice cultivation system. The innovative rice cultivation system involves a combination of four basic elements: transplanting young seedlings at low density with spacing that allows more roots to grow in less water (and to save the number of seedlings to be transplanted); enriching the soil with organic matter (to improve its capacity to retain nutrients and water, but also to increase the microbial life of the soil and provide a good substrate for the development of the roots); reducing water inputs and controlling the good growth of the plant by drying it out to allow aerobic growth (which can only occur when in contact with air), thus promoting growth without the need to add fertilizers (and especially chemical fertilizers). Thus, without the addition of synthetic plant protection products, the innovative rice cultivation system (which requires more careful work and know-how) promises a good harvest with higher yields, and since the rice field is not permanently flooded, methane emissions are reduced.

The innovative rice cultivation system technique sometimes struggles to develop and become more widespread because of the knowledge required for cultivation practice, particularly with regard to the issue of water control, which is not always possible in monsoon regions and, when possible, is demanding in terms of observations, technical handling and work time. Therefore, even when the innovative rice cultivation system technique is possible, it can be demotivating. Training motivated farmers is therefore a challenge in itself, but one that responds well to the issues at stake: societal demand for sustainable development and quality products; the need to repair soils, clean up pollution and deal with water scarcity and global warming.

A second innovative technique consists of combining rice cultivation and livestock, sometimes in a marginal but effective way. This is the case with rice cultivation, where approximately 20 ducks are kept on each small fenced plot. There are many advantages: the ducks' droppings fertilize the soil, their light trampling aerates the soil and the ducks eat insects or crabs in the rice field and lay eggs. As a result of these endogenous and natural inputs, rice cultivation offers interesting yields (without the expense of chemicals).

Moreover, replanting trees around the rice field allows stabilization of the water table with the support of NGOs or companies (e.g. Pur Projet, see: wwwpurprojet.com), whose objective is to encourage carbon offsets by replanting trees.

As part of the principle of the *Sufficiency Economy* promoted by the Thai monarchy, Pur Projet has replanted at least 100,000 trees in the country, which is certainly far from compensating for deforestation around the world, but the initiative marks a notable inflection point.

The main obstacle to these creative practices, in addition to the training of mobilized farmers, is the *Green Washing* linked to the bypassing of the respect of the specifications by displaying green on the commercial label, but continuing to use synthetic phytosanitary products with ease and commercial cunning, but also probably and simultaneously under the pressure of the agro-industry unfavorable to creative cultivation practices that do not propose hybrid seeds or synthetic phytosanitary products (both being available for sale from the agro-industry and its suppliers). Lobbies are present in Asia, particularly in Thailand, which to date has been a significant user of chemical inputs.

The commercial logic (to the detriment of a policy favoring local authenticity and control by the farmers) by the economic dependence it entails, also generates, correlatively, in Asia or Africa, the business of counterfeit products from which industrial seeds and synthetic phytosanitary products do not escape. Traceability is therefore fundamental, but it can be controlled at the local level by practices that respect the Natural Environment and use local products and organic inputs.

This said, rice cultivation that is more respectful of the Natural Environment (and ideally organic) could gradually impose itself (even against agro-industrialists) due to the deep aspirations of the populations who are fed up with pollution and the impacts on human health, and the useless sacrifice of the Natural Environment, biodiversity and ecosystems. The government of a country can enlighten the choice of private interests and lobbies in the name of collective interest and sustainable development at a historical moment when global warming is an unavoidable challenge. In a sustainable model, technology has its place but is free from marketing injunctions.

2.4. Conclusion

The exploitation of resources is exemplary if it does not generate any waste, but allows a total valorization, such as the coconut tree and its fruit in India or Thailand, but should they be exceptions or a value chain model (CtoC type) to follow?

Global warming and the need to take sustainable development into account, which are emerging as new and increasingly avoidable constraints for manufacturers, must lead to public and private strategic reorientations by putting innovation at the heart of the reflection on the accuracy and relevance of its use as support for an agriculture that must be sustainable.

The classical pattern implemented since the beginning of the European Industrial Revolution concerning resource depletion, negative externalities (including pollution and waste) and inducing global warming is obsolete. Moreover, techno-marketing generates an unbridled race to the front, making us lose sight of the crucial issues concerning the raison d'être of technology. This is why EFI, which in itself is neither a brake on technological development nor a call for a return to tradition, is based on organizational innovation, innovative creativity (including bricolage and improvisation), respect for the Natural Environment, the development of good practices and the design of useful and adapted technological products without wasting resources.

Within communities of practice, exchanging good practices, sharing knowledge and know-how, and encouraging learning mobilize bricolage and improvisation to adapt to the context of the place and the moment by doing things better in other ways.

Concerning agriculture, in particular, it is no longer possible to allow logics based on increasing needs for resources to continue, generating major collateral effects such as the reduction of biodiversity, the disappearance of ecosystems and massive deforestation.

Humans have the ability to choose their model and respect the Natural Environment. Technology can be a precious help to achieve this if we select a satisfactory technological level. The strong constraint of global warming guides this choice.

In the era of "Makers" and "FabLabs", communities of practice, cocreation and participative open initiatives, the technology is decentralizable and adaptable to the local needs of farmers concerned with respecting the Natural Environment.

States or international organizations (the United Nations Organization (UNO) including its branch for food and agriculture: the Food and Agriculture Organization (FAO)) must encourage local initiatives and not deceive themselves on the stakes by importing unsuitable and counterproductive models. Promoting food for all and good nutrition in the framework of sustainable development is a priority that requires education and innovative creation within local communities of practice. The important thing is to encourage, within the framework of sustainable development, the creation of value chains through local valorization of local products by farmers, but also a local agri-food industry.

2.5. References

Anderson, C. (2012). *Makers: The New Industrial Revolution*. Crown Business Book, New York.

Aune, J.B., Doumbia, M., Berthe, A. (2007). Microfertilizing sorghum and pearl millet in Mali: Agronomic, economic and social feasibility. *Outlook on Agriculture*, 36(3), 199–203.

Basquin, H., Charhon, F., Cissokho, M., Le Gauyer, G., Rouillé d'Orfeuil, H., Vielajus, J.-L. (2014). *Nourrir les villes : défi de l'agriculture familiale*. CFSI and Fondation de France, Paris.

Benyus, J. (1997). *Biomimicry, Innovation Inspired by Nature*. HarperCollins, New York.

Beucher, O. and Bazin, F. (2012). *L'agriculture en Afrique face aux défis du changement climatique*. Organisation Internationale de la Francophonie, Quebec.

Bouvier-Patron, P. (2020). Stratégies d'entreprise et innovation frugale environnementale : proposition d'une matrice stratégique. *Recherches en sciences de gestion*, 136(1), 39–65.

Bouvier-Patron, P. (2021a). Co-creation and innovation: Strategic issues for the company. In *Innovation Economics, Engineering and Management Handbook 2*, Uzunidis, D., Kasmi, F., Adatto, L. (eds). ISTE Ltd, London, and John Wiley & Sons, New York.

Bouvier-Patron, P. (2021b). From improvisation to innovation: The key role of "bricolage". In *Innovation Economics, Engineering and Management Handbook 2*, Uzunidis, D., Kasmi, F., Adatto, L. (eds). ISTE Ltd, London, and John Wiley & Sons, New York.

Bouvier-Patron, P. (2023). *Frugal Innovation and Innovative Creation*, ISTE Ltd, London, and John Wiley & Sons, New York.

Braungart, M. and McDonough, W. (2002). *Cradle to Cradle: Remaking the Way We Make Things*. North Point Press, New York.

Brown, J. and Duguid, P. (1991). Organizational learning and communities of practice: Towards a unified view of working, learning and innovation. *Organization Science*, 2, 40–57.

Cessac, M. (2019). Agriculture : des semenciers qui s'enracinent. *Jeune Afrique* [Online]. Available at: https://www.jeuneafrique.com/mag/821924/economie/agriculture-des-semenciers-qui-senracinent/.

CIRAD (2013). Gérer collectivement la biodiversité agricole en Afrique. *Valorisation et innovation en partenariat*, 24, January [Online]. Available at: http://www.cirad.fr/var/cirad/storage/original/application/18b46b74ad049e02337fd52ac5c74786.pdf.

Clavel, D. (2016). Afrique : une politique semencière sous influence. *Natures Sciences Sociétés*, 24(2), 168–172.

Cruz, J.F., Hounhouigan, D.J., Havard, M., Ferre, T. (2019). *La transformation des grains*. Collection Agricultures Tropicales en Poche, Quæ, Presses Agronomiques de Gembloux, CTA, Versailles, Gembloux, Wageningen.

Cunha, M., Rego, A., Oliveira, P., Habib, N. (2014). Product innovation in resource-poor environments: Three research streams. *Journal of Product Innovation Management*, 31(2), 202–210.

FAO (2020). *La situation mondiale de l'alimentation et de l'agriculture 2020 : relever le défi de l'eau dans l'agriculture*. FAO, Rome.

Fusiller, J.-L. (1994). Afrique de l'Ouest et du Centre : maïs ou sorgho ? *Agriculture et Développement*, 2, 20–28.

Gershenfeld, N. (2005). *FabLab: The Coming Revolution on Your Desktop – From Personal Computers to Personal Fabrication*. Basic Books, New York.

Jacquemot, P. (2021). *Souverainetés agricole et alimentaire en Afrique, la reconquête*. L'Harmattan, Paris.

Jacquemot, P. (2022). Afrique, des campagnes et des villes – Des systèmes agroalimentaires territorialisé [Online]. Available at: https://www.willagri.com/2022/10/03/afrique-des-campagnes-et-des-villes-des-systemes-agroalimentaires-territorialises/.

Jones, C. and Pimdee, P. (2017). Innovative ideas: Thailand 4.0 and the fourth industrial revolution. *Asian International Journal of Social Sciences*, 17(1), 4–32.

La Croix (2019). Arroser moins pour récolter plus : le riz thaï qui défie la sécheresse. *La Croix Numérique* [Online]. Available at: https://www.la-croix.com/Economie/Arroser-moins-recolter-riz-thai-defie-secheresse-2019-12-12-1301066026.

Maslow, A.H. (1943). A theory of human motivation. *Psychological Review*, 50(4), 370–396.

Ministère de l'Agriculture et de l'Alimentation Français (2019). Fiche Thaïlande [Online]. Available at: http://agriculture.gouv.fr/politiques-agricoles-fiches-pays.

Phélinas, P. (2010). Le marché, l'état et la compétitivité du riz Thaïlandais. *Annales de géographie*, 1(671–672), 156–173.

Portères, R. (1955). L'introduction du maïs en Afrique. *Journal d'agriculture tropicale et de botanique appliquée*, 2(5–6), 221–231.

Prahalad, C. (2005). *The Fortune at the Base of the Pyramid: Eradicating Poverty through Profits*. Wharton School Publishing, Philadelphia.

Roos, E. (1985). Dégradation des terres et développement en Afrique de l'Ouest. *Bulletin de recherche agronomique Gembloux,* 20(3/4), 505–537.

Seyfang, G. and Smith, A. (2007). Grassroots innovations for sustainable development: Towards a new research and policy agenda. *Environmental Politics*, 16(4), 584–603.

3

The National Innovation System as Applied to Agriculture: A Methodological Proposal for Diagnosis in Africa

3.1. Introduction

Since the late 1990s, the "market model" has begun to lose its legitimacy as the dominant regulatory framework for development (Compaoré 2015; Sawadogo and Sawadogo/Compaoré 2018). Indeed, the World Bank's 1998 development report was entitled *Knowledge for Development* (World Bank 1998); the following year, UNESCO's report was entitled *Science in the Twenty-first Century, A New Commitment*. For any informed observer of development issues, it is clear that the Washington Consensus, which enshrined the market model, was in crisis and that a new consensus was emerging (Cimoli et al. 2009). Thus, the new millennium is beginning in Africa with the deployment of a new development model that now places a premium on innovation. This is a model of knowledge-based development which, since 1998, has given rise to several meetings on the African continent. In 1999, during the year of publication of the abovementioned UNESCO report, the *Knowledge Management in Africa* (KMA) network organized a symposium in South Africa on the theme *Knowledge to Address Africa's Development Challenges*. Two years later, the creation of the New Partnership for Africa's Development (NEPAD) set out to create new conditions for development in Africa. In 2003, the African Ministerial Council on Science and Technology at its first meeting in Johannesburg, South Africa, reaffirmed the role of science in the development of the African continent. In January 2007, this process was crowned by the Addis Ababa Declaration on Science, Technology, and Scientific Research for Development at the African Union Summit. From then on, in all sectors, and in agriculture in particular,

Chapter written by Eveline M.F.W. SAWADOGO/COMPAORÉ and Natewindé SAWADOGO.

new concepts began to permeate discourse and practice, notably through the use of the national innovation system (NIS) framework. This framework, developed by Freeman and Lundvall (1988), is a concomitant model of analysis and intervention that makes innovation the driving force of economic development. The NIS has been tested since the early 2000s. It has become very popular in public policy and research contexts in Africa (Muchie et al. 2003; Touzard et al. 2015).

However, the dissemination of this new development model in agriculture remains a challenge for actors (researchers, development practitioners and politicians) in the sector. Indeed, the linear and diffusionist conception of innovation has deeply marked the teaching structures, public governance and cognitive referentials without notable positive results with regard to current development indicators. At the beginning of the 21st century, the search for an alternative remained more at the level of "expressions of interest" than concrete proposals for methods and practical tools to analyze an agricultural innovation system (AIS) (OECD 2013) and its transformation for development. The words of a French agronomist in the late 1990s are quite illustrative:

> Interventions by public authorities [...] will undoubtedly remain necessary for a long time to come. However, it is now necessary to completely overhaul the approach and methods relating to the design, formulation, implementation, monitoring and evaluation of projects, drawing lessons from past experience. Above all, it is important to design democratic procedures that will make it possible to truly take into account the respective interests of the various social classes, strata and categories [...] (Dufumier 1996, p. 10).

These statements call for a shift from a top-down development approach to a participatory and democratic development approach. They refer to the state of mind of this period, proposing the main variables to be taken into account for the evaluation of agricultural projects for agricultural development in Africa, in particular. Additionally, other authors have addressed these issues since the late 1990s by mobilizing the theoretical assumptions that structure the NIS analysis framework by developing sectoral methodological guides for understanding AISs (OECD 2013). The remaining challenge is to develop a methodological guide for diagnosing AISs in Africa to reveal how knowledge can become the main driver of socioeconomic development. This chapter analyzes how past development approaches and the application of the NIS to the agricultural sector can be used to propose an integrated diagnostic methodology. It draws on the work on techno-economic networks (Callon 1991) and contributes to the more global work (Casadella and Benlahcen 2006) through a documentation of African experiences.

The chapter is structured as follows. The first part presents the approach used. The second part reviews the approaches to agricultural innovation in Africa based on the experience of Burkina Faso. Finally, the third part makes a methodological proposal for the diagnosis of AISs in Africa.

3.2. Methodology

The development of the methodological proposal is based on two strategies in the context of the study in Burkina Faso. The first strategy involves a literature review on linear approaches to agricultural development.

It is based on a description of a case study analysis of agricultural development interventions using linear approaches. It describes these interventions in concrete terms and identifies their theoretical foundations in particular. These approaches are approached as "technologies", that is, "social practices and social relations" (Sharif 2004, p. 89). The chapter proposes to examine at the same time the transformations of these technologies in relation to the contexts of their use. This exploration is complemented by an analysis of the literature concerning work carried out in other situations outside Burkina Faso. This method thus aims to diagnose the strengths and weaknesses of these approaches to confront them with interactional approaches in the process of developing the diagnostic methodology.

The second strategy is to examine case studies of interactional approaches to agricultural innovation. This work is based on cases that have involved the authors of this chapter. They are based on nearly 10 years of practical experience in rural development support and scientific work on the NIS and its application in the agricultural sector known as the AIS in Burkina Faso. The first case study from 2010 concerns the biotech cotton innovation system in Burkina Faso, which was explored as part of a Master's degree in Science, Technology and Society (Compaoré 2009; Sawadogo/Compaoré and Sawadogo 2018). This research thus studied the structure and dynamics of the actors involved in the system as a means to understand the process of translating new technologies into the cotton sector and to assess the outcomes in terms of economic and social development benefits for the actors involved. The second study is based on 3 years of doctoral research (2011–2014) on the conditions of use of the NIS model in development planning and its translation into the Burkinabé agricultural sector (Compaoré 2015; Sawadogo/Compaoré 2017). The objective was to investigate the facilitative role of the NIS framework in facilitating socioeconomic development in Burkina Faso through two levels of dissemination, the policy (macro) and operational (micro) levels. This study expanded on previous research on the adoption and implementation of the biotech cotton innovation. In addition to these two case studies, additional research and interventions on specific dimensions of the AIS or the overall NIS were conducted in three projects. In 2016, both authors participated in the implementation of the joint FAO Government of Burkina Faso–Agrinatura project on capacity development of AISs, which aimed to support the development of policies and approaches in the three continents: Asia/Africa/Latin America (Sawadogo/Compaoré 2017). Over the period 2015–2018, one of the authors led a 3-year project (2015–2018) with other colleagues (IDRC-DID-INERA and Université de Laval) on "the diffusion of financial services and innovation among rural populations in Burkina Faso". Rural finance is considered to be an essential tool for poverty reduction. The objective was to analyze the link between financial resources and the diffusion of innovation. This research-driven project targeted farming communities (Pô, Sapouy, Gourci and Yako) and assessed the impact of traditional community financial services on agricultural production, food security and resilience, as well as informal lending, borrowing and savings credit systems (Sawadogo/Compaoré 2017). However, a better way to include smallholder farmers has yet to be demonstrated. In 2016, one of the authors had the privilege of serving on the national working group on agricultural technology needs assessment as part of the Poznan Global Strategic Program on Technology Innovation Potential and Climate Change Adaptation. In 2017, a project to assess the gender gap in agricultural productivity assessment was conducted in seven countries (Nigeria, Ghana, Burkina Faso, Mali, Uganda, Tanzania and Kenya). We also explored the relationship between innovation and development in a contribution to the ownership of knowledge-centered development in Africa and its challenges (Sawadogo and Sawadogo/Compaoré 2018). This research series articulates work on the broad theoretical understanding of the NIS model with more specific concrete interventions in the AIS by applying the NIS model to the agricultural sector. It not only feeds a theoretical mastery of the experiences, properties and dynamics of the AIS in Burkina Faso but also explores other situations in Africa. It leads to the mobilization of this experience to propose a methodology for diagnosing AISs adapted to the specified geographical context. The methodological framework proposed here is not intended to be universal. Its modest objective is to contribute to the knowledge of the AIS of Burkina Faso, in particular, and those of African countries that share more or less the same structural characteristics. It thus invites reflections in collaboration with other African colleagues interested in the same issues; and why not encourage more global research partnerships with the aim of developing a more robust methodology for the diagnosis of AIS in Africa. All of these case studies, works and interventions are analyzed from the point of view of their strengths and weaknesses in their ability to activate interactional models in the realization of agricultural innovation. This set is based on the knowledge acquired sometimes in case studies that refer to linear models and in the NIS framework. The objective, although contextualized to a single country, is to develop an evidence-based methodological approach with the ambition of constituting a more global reference framework for other researchers.

3.3. Linear model of agricultural innovation

In Burkina Faso, "the integrated approach to cash crops" is the very first approach to agricultural technology transfer. The period from 1900 to the 1970s was marked by what is known as technology transfer from developed to underdeveloped countries (Gaillard 1999). This integrated approach, when used as a tool for the diffusion of innovation, consists of a process that extends from production to processing and marketing of derived products on the market. In the agricultural sector, where this tool has been mainly applied, vertical integration empowers industries by giving them the role of mediators of innovation, as well as suppliers of inputs and credit, etc. They provide farmers with everything they need for their cash crops, and the farmers have to pay them back after the harvest. The research institutes, which conduct research to solve the problems associated with these cash crops, then interact only with these industries and not with the farmers (Belloncle 1987). Researchers in this linear process are informed of farmers' needs by technical support agents employed by the industries. There is no direct and simultaneous contact between the agricultural agents, the farmers and the researchers (Evanson 1987). The researcher is in his expert world, leaving the producer in the hands of the agricultural technical agent who is the intermediary between the researchers and the producers. This is an approach with a top-down communication model, leaving little room for the participation of the various actors. The following table gives the characteristics of this approach.

	Components			**Potential for participation**	**Wealth creation from knowledge**[1]					
Diffusion of the innovation approach	**Role of the government**	**Role of universities**	**Role of industries**		**1**	**2**	**3**	**4**	**5**	**6**
Top-down or downward			X	No participation		X				

[1] **1:** Entrepreneurs who are trying to create new and innovative businesses that deploy their knowledge in new ways. **2:** Business leaders seeking to improve productivity through the effective application of new knowledge and concerned with the "reinvention" of the business through a form of entrepreneurship. **3:** Leaders of R&D structures in the public and private sectors attempting to improve the use of knowledge created within their groups. **4:** Knowledge-intensive business service providers trying to create a business by enabling connections and flows within the knowledge system. **5:** Government policy agents at the local, regional and national levels charged with improving the effectiveness and efficiency of the innovation system. **6:** Supply chain "owners" wanting to improve the efficiency/effectiveness of their system-level networks through innovation

Table 3.1. *Description of the linear approach to agricultural innovation*

On the theoretical level, the conceptual bases of such a model are well known. Indeed, predominant in the 1950s, 1960s and early 1970s, it was through linear diffusionist theories that the process of economic growth was studied. These theories generally considered innovation to be dependent on scientific discoveries. According to Godin (2009), the first model was called Science-Push (Godin 2009). It was then followed by the demand-pull model. In the first one, it is assumed that any scientific discovery necessarily leads to technological innovation through its use by "knowledge users", particularly companies. The same posture is adopted in the second one, which attributes the development of the discovery to the formation of demand and the development of the market (Manley 2003; Godin 2009).

3.4. Interactional models of agricultural innovation

Toward the end of the 1970s, the lack of satisfaction with development experiences challenged previous models and began to transform development practices. This led to the development of an increasing number of interactional approaches.

3.4.1. *Participatory approach to agricultural development or "training and visits" approach*

The participatory approach to innovation diffusion, called *top-down bottom-up*, was developed by Daniel Benor for the World Bank in the 1970s (Russell 1987). It consists of organizing visits to farmers by technical agents at a time convenient to the agent and the farmer. One person who lived through this period describes the approach as follows:

> In the late 1980s, a World Bank consultant, Benor, developed a system called TV (training and visit). However, it was still the same linear process where there was no communication between researchers and farmers. The relationship is between researchers and technical agents who always meet. The farmer is kept away, and only the technical agent can meet the researchers. So the linear thinking still exists, but there was a dialog between technical agents and researchers during meetings and a dialog between these technical agents and the farmers [...] (interview 16).

Thus, it is still a producer-oriented approach, but with a top-down ascending communication model, leaving more room for producer participation. Table 3.2 gives the characteristics of the approach.

	Components			Potential for participation	Wealth creation from knowledge[1]					
Diffusion of the innovation approach	**Role of the government**	**Role of universities**	**Role of industries**		**1**	**2**	**3**	**4**	**5**	**6**
Top-down bottom-up (Training and Visit)			X	Limited participation		X				

[1]Same as in Table 3.1.

Table 3.2. *Description of the training and visit approach*

3.4.2. *In search of an interactional innovation diffusion approach*

This is the farming systems approach. In this interactional approach, the goal is that "specific groups with shared needs living in a defined geographical area actively seek to identify their needs, make decisions, and establish mechanisms to meet their needs" (Rifkin et al. 1988, as cited in Bjaras et al. 1991). It is an approach that is strategic in orientation and goes beyond the producer with an interactional communication model, allowing all actors to interact with each other directly. Table 3.3 gives the characteristics of the approach.

	Components			Potential for participation	Wealth creation from knowledge[1]					
Diffusion of the innovation approach	**Role of the government**	**Role of universities**	**Role of industries**		**1**	**2**	**3**	**4**	**5**	**6**
Agrarian or agricultural system			X	Limited participation		X		X		

[1]Same as in Table 3.1.

Table 3.3. *Description of the farming system approach*

In 2005, two experiments involving interactional approaches emerged. One is based on the NIS analysis framework that is the subject of this chapter, and the other on the establishment of multistakeholder platforms (MSPs). According to the Ministry of Research and Innovation of Burkina (2012, p. 11), the MSP is "a grouping of means (essentially material and human) for a specific open community where users, mainly companies, interact (equipment localization, service experimentation), allowing them to carry out their research and development (R&D) innovation projects".

	Components			**Potential for participation**	**Wealth creation from knowledge**[1]					
Diffusion of the innovation approach	**Role of the government**	**Role of the universities**	**Role of industries**		**1**	**2**	**3**	**4**	**5**	**6**
Multistakeholder platform			X	Increased participation		X	X	X		X

[1] Same as in Table 3.1.

Table 3.4. *Description of the multiactor platform approach to agricultural innovation*

Since the OECD's landmark publication on the management of NISs in 1999, the framework called NIS has become increasingly popular. It was first developed as a theoretical framework, concept and analytical perspective and then used to study the flow of technology and information between individuals, firms and institutions. This systemic analysis framework diverges from previous models characterized as linear.

3.4.3. *Description of the framework of NISs*

Origin and pioneering work

The use of systemic analysis applied to innovation is at the origin of the NIS concept, which structures the interdependencies between basic sciences, marketing, manufacturing and research and development (R&D) (Manley 2003; Godin 2009). Innovation is seen as an interactive and complex process. The concept of NISs was developed in the 1980s by Freeman (1982) and Lundvall (1985) to explain how technological innovation results from the complex interaction between institutions (Lundvall 2007). This conceptual framework was then adopted in science, technology and innovation studies from the 1990s onward to study innovation systems (Lundvall 1992; Nelson 1993; Freeman 1995; Lundvall 2007). Godin (2009) has shown that the "systems approach" developed by the OECD is largely based on the abovementioned research work since the 1960s.

Objectives/functions of an NIS

According to Lundvall, the objective of an NIS is to achieve international competitiveness of national economies and to lead to the growth of the national economy. For him, there seems to be a broader social acceptance of the concept of "national". However, this is not the only relevant level to consider (Lundvall 2007). It follows that the social dimension remains uncertain to be realized within the national framework. Rather, it is through international (continental) organizations, on the one hand, and global organizations (e.g., United Nations agencies, global environmental organizations), on the other hand, that this dimension could be realized. In the case of international organizations, politicians and experts at this level are more oriented toward strengthening the economic growth of their respective regions and preventing international conflicts within the community of the countries they represent (Lundvall 2007). In the case of global organizations, it is becoming more evident to experts and political actors that the long-term survival of the global economy depends on ecological sustainability and a reduction in extreme social inequality at the global level (Lundvall 2007). Clearly, the NIS remains limited in regard to promoting or addressing the social dimension of development.

The nature of complex and interactive innovation processes

An NIS requires *knowledge producers,* in this case universities. Universities are essential to the creation or invention of any innovation, as the performance of the system depends on human capital and capabilities. These knowledge producers are produced by universities, technical organizations and schools (e.g. Lundvall 2007). Universities in particular have their own research units that facilitate market formation by providing new knowledge and skills and conducting research (Lundvall 1985, 2007; Freeman 1987; OECD 1999). These knowledge-producing institutes are centers of basic research and scientific training (Lundvall 1985; OECD 1999). In other words, universities as knowledge producers engage in basic science and technological research. At the same time, universities provide scientists and technologists with higher education that enables them to generate the innovation that businesses and governments need. Nelson has also emphasized this role of universities: "[Universities provide] higher levels of scientific training and management skills to coordinate what is inevitably a multi-person or multi-firm affair" (Nelson 1993; Feinson 2003, p. 24). However, in the NIS model, this is not enough; other actors are needed.

The second key actor is *companies*. They are all considered to be economic actors that use the innovation created by knowledge producers such as universities and research centers (OECD 1999). In NIS, firms contribute to the diffusion of knowledge in the market (OECD 1999). Lundvall (1992), Nelson (1993) and Edquist (1997) have examined the idea of a system where firms interact with the two variables mentioned above. As demonstrated by the OECD (1999), firms or industry do not operate alone, as governments, such as universities, are also an integral part of the system in the innovation process. Firms, on the other hand, are seen as economic actors whose role is to conduct R&D activities to develop products for commercialization. They launch various innovative products and exploit new scientific discoveries (OECD 1999). For the authors of the NIS framework, limiting themselves to firms means staying within the "demand-pull" model. Both firms and universities can only work properly if there is an enabling environment. Indeed, many innovation system scholars have explicitly explained that the government should be actively involved (Freeman 1995) in promoting innovation. During the 1990s and 2000s, the importance of the innovation system in relation to new technologies became widely accepted, and its use began to spread in both developed and developing countries (OECD 1999). In addition, the importance of knowledge in the process of innovation diffusion has received particular attention. As a result, knowledge has been identified as an engine (1999) for achieving and enhancing sustainable economic growth. This contemporary vision has been widely adopted and has had a significant impact on innovation strategy and policy, particularly in developed countries. The contributions of individual institutions are not considered in isolation but with "how they interact with each other as elements of a collective system of knowledge creation and use, and their interaction with social institutions: values, norms, legal frameworks, etc." (Smith in OECD 1999, p. 24). In this interaction, the government can play a key role through its policy options and regulations. Innovation systems approaches, underpinned by assumptions and framing conventions based on the experiences of developed countries, have been applied in developing countries with little modification from the different contexts.

The importance of knowledge and technology flows

In general, the theory of NISs emphasizes that the flow of technology and information between individuals, firms and institutions is essential to the innovation process. Thus, innovation and technological development result from a complex set of relationships between actors in the system, which includes firms, universities and government research institutes (Godin 2009). The inclusion of the country's research system in this complex set of relationships is another distinctive feature of this approach. Godin notes that:

> The NIS framework suggests that the ultimate goal of the research system is innovation and that the system is part of a larger system composed of sectors such as government, universities, industry and their environment. The framework also emphasized the relationships between components or sectors as a "cause" explaining the performance of innovation systems (Godin 2009, pp. 476–477).

This theory postulates that understanding the links between the actors involved in innovation is the key to improving technological performance. Innovation and technical progress result from a complex set of relationships between actors producing, disseminating and applying various types of knowledge (OECD 1999). It is seen as a new way of doing things or a redesign of an existing product for a "niche" market or a change in the way work is done in an organization (OECD 1997).

Role of institutions and interactions between institutions

The NIS concept also involves two main categories: market institutions and non-market institutions (Manley 2003). Both are considered for their potential to influence the direction and speed of innovation and technology diffusion in a country (OECD 1999). Importantly, this theory was developed primarily from the study of innovation systems in developed countries (Freeman 1987; OECD 1999). In this regard, as Sharif puts it, "the NIS is seen as the set of institutions and practices that interact to produce and diffuse new technologies" (Sharif 2006, p. 84). Here, the NIS is seen as an object of study in the sense of calling the British NIS, the US NIS, the Burkina Faso NIS, etc., a system that can be studied as an object or transformed as matter. As Freeman puts it, "[...] the network of public and private sector institutions whose activities and interactions initiate, import, modify and diffuse new technologies" (Freeman 1987 in OECD 1999, p. 10).

3.4.3.6. *Role of government and the national policy framework and the nation state*

Technology acquisition, use, and diffusion involve the government, whose role in the system is critical. Through its policymakers, the government plays a key role in making choices that inform strategies at the national level (Freeman 1987). Indeed, the government is seen as a key actor influencing foreign direct investment (FDI) because "despite their limited size, FDI inflows have had a positive impact, while generating employment in the formal sector and generating local value added" (United Nations 2009, p. 2). On the other hand, such a flow is supported within the framework of government regulation of IPR (Intellectual Property Rights) in which the processes are decomposed into three stages: innovation process, commercialization process and IPR protection process. By designing the IPR system in the interest of direct scientific research by business and government, such as funding universities and supporting business research and development (R&D), "it is the government's responsibility to contribute to the formation of the human and social capital needed to evaluate, select, implement and modify foreign technologies" (Feinson 2003, p. 23). R&D can be used for knowledge assimilation and domestic technological creation, as Freeman (1987) shows in his case study of Japan. Incentives created by the government are needed to define these aspects at the level of national policy orientation to provide administrative support as well as facilities and equipment to strengthen and legitimize the networking of knowledge producers and users (Lundvall 1985; Freeman 1997; OECD 1999). Government commitment as well as specific actions to provide actors with such a context are therefore necessary to create a social and political context that is conducive to the networking of firms and knowledge producers. In this model, the relevant actors and their roles are better known, and the effects of their combinations are controlled. However, although the actors interact, the econometric treatment of the model does not allow us to understand the processes through which the actors interact. Innovation is, however, a social process that needs the movement of actors whose parameters must be taken into account in any analysis.

The NIS as a strategic analysis and diagnostic tool

Since the popularization of the concept by the OECD, NISs have been used as a policy tool or model to guide innovation systems at the national level. The government's interest in the policy tools of the NIS is particularly relevant to this chapter, as it refers to a third definition of the framework in which, with respect to the policy circle, the NIS is seen as a "policy tool" (Lundvall 2007, p. 1). This concept then becomes a steering tool to give practical coherence to decision makers in order to design and implement their policies. It can thus be seen as a policy narrative through which decision-makers develop their policies. Godin clearly captures this aspect when he says that "decision-makers elaborate their problem through conceptual frameworks that structure political action" (Godin 2009, p. 5).

It is in this respect that the NIS can be seen as a model, that is, "a technology that allows us to [...] make things work in the world" (Morrison and Morgan 1999, p. 32). In this chapter, this dimension (use of the NIS framework as a policy tool in Burkina Faso and a sector development strategy) is considered. In the specific case of Burkina Faso, a project was thus initiated in 2006 by FRSIT (representing the government) and IDRC, with the aim of introducing the NIS framework as a policy tool to improve what might be called the Burkina Faso NIS.

	Components			Potential for participation	Wealth creation from knowledge[1]					
Diffusion of innovation approach	**Role of the government**	**Role of universities**	**Role of industries**		**1**	**2**	**3**	**4**	**5**	**6**
National innovation system	X	X	X	Participatory methods but not adapted to all contexts, especially in developing countries	X	X	X	X	X	X

[1] Same as in Table 3.1.

Table 3.5. *Description of the NIS approach*

3.5. Beyond the NIS: taking context as a key aspect

The sociological analysis perspective developed by the French school of innovation, in terms of the "Techno-Economic Network" (TEN), provides a model that complements the shortcomings noted in the different approaches described above. Its less positivist approach allows for the possibility of adapting the modalities of communication in the NIS. Indeed, according to Callon et al. (1992, p. 215), a techno-economic network (TEN) is "a coordinated set of heterogeneous actors (public laboratories, technological research centers, industrial firms, users) that collectively participate in the development and diffusion of innovation and, through numerous interactions, organize the relations between scientific and technological research and the market". This model makes it possible to integrate the strengths of previous approaches, whose relevance has been finalized in the context of industrial countries. In terms of communication, the tools of communication with a predominantly illiterate population have stabilized. This is what is missing from the NIS model, which in fact bears the characteristics of its context of development, which is that of developed countries (Katz et al. 2014). The proposed practical methodology for diagnosing AISs takes into account the strengths and weaknesses of the main agricultural innovation approaches used thus far in Burkina Faso and in Africa in general.

3.6. Practical methodology for the diagnosis of AISs in Africa

Based on the above developments, we identify six key steps relevant to the diagnosis of AISs in Africa.

Step 1: understanding the key characteristics of an AIS

Variables	Data collection techniques
– Diversity of sectors (agricultural industry, agricultural research and education systems, agricultural government agencies) – Diversity of "actors" (farm businesses, research and educational institutions, agricultural financial institutions, farm workers, standards) – Variety of actors (different categories of agricultural enterprises, different categories of research and educational institutions, different categories of financial institutions, different socioprofessional categories, standards) – Existence of links between sectors and between actors – Existence of interactions between sectors and between actors – Complementarity between sectors and between actors – Dynamic: it evolves over time and according to the issues – Describe the main functions of an effective and efficient agricultural innovation system – Enable knowledge flow – Facilitation (support: institutional, financial, expertise, etc.) – Communication – Collective action – Generating innovation – It sets standards for effective and efficient performance	Quantitative and qualitative

Step 2: identifying the key elements of an AIS. What types of groups, networks, would be needed to create an effective (optimally desired) AIS?

Variables	Data collection techniques
Data to be collected I: consist of a techno-economic analysis and include first, the identification of the actors, notably (i) public laboratories, (ii) technical research centers, (iii) industrial companies, (iv) financial organizations, (v), users and (vi) public authorities. Secondly, the characterization of these actors of the techno-economic network according to their corresponding centers: (i) Scientific center (S), referring to the production of certified scientific knowledge (journal/laboratory articles), reports, working notes (with limited dissemination, even semiautomatic-confidential), instruments (technical objects), incorporated skills (researchers trained in universities in transition to the industry/development sector considered). (ii) Technological center (T) consisting of the identification of material objects designed with a coherence of their own. (iii) Market center (M) consisting of the identification of distribution and commercialization networks of organizations (private, public; commercial/non-commercial), as well as users or groups of consumers linked to distribution and redistribution organizations. To interact and justify the levels of interaction, other actors are needed: the intermediaries. *Data to be collected II*: collection of written documents (scientific articles, reports, patents, etc.), incorporated competences (researchers changing laboratories, engineers moving from one company to another, etc.), (financing or source of income) (cooperation contracts, financial loans, purchase of goods or services by customers, etc.), more or less technical objects (prototypes, machines, products intended for financial consumption, etc., test sites, standards, rules of the art, methods). These will be channeled through the two intermediate centers: transfer center (TC) for the organizations of scientific and technological interactions and development center (DC) for the production and distribution activities between technology and market (technological resources and market).	Data collection techniques include questionnaires, literature review, including archives, interviews and direct observation. Data will be analyzed using bibliometric and scientometric techniques, as well as content analysis.

Step 3: assessing the existing AIS

Research question: what are the strategic goals of the national AIS?

Variables	Data collection techniques
Data to be collected: The first step will be to identify the assignments, consisting of: i) Elucidating general principles of action/policy. ii) Defining the framework for action/policy. Secondly, it will identify the strategic "products" concerned by the national policy on AIS, namely: a) lists of products or product families to be developed as strategic for the political missions; b) the areas of development related to these products that are considered strategic for the political mission.	A literature review and interviews will be used for data collection. Content analysis and related word analysis will be used. *Analysis results*: identification of criteria used for policy options; lists of products, or product families, identified by the policy, and associated rationale.

Step 4: assessing the applied/implemented model for developing an AIS

This step assesses the epistemological and theoretical bases underlying the interventions on the national AIS. The question to be answered is whether the agricultural innovation under consideration is part of a linear approach to the diffusion of innovation or whether it is interactional.

Step 5: comparison of the existing AIS and the desired optimal AIS (standard)

Research question: what has been achieved since then?

Variables	Data collection techniques
Data to be collected: Techno-economic network models through descriptions and desired comparative analysis against existing AIS provided by the techno-economic assessment. This will identify configurations that the techno-economic network may adopt throughout the implementation of AIS interventions.	Quantitative and qualitative combining the quantitative and qualitative tools from steps 1–4.

Step 6: interpretation of the AIS portfolio (outputs/outcomes)

Variables	Data collection techniques
The analysis will consist of tracing the dynamics of the evolution of TENs; and identifying the types of networks: incomplete network, "when one or more categories of actors that constitute it [...] are almost totally absent or underdeveloped"; chained network, "If [...] all actors and centers are in position and strongly structured"; establish the degree of network integration by analyzing its "dispersion", if its density is low or high, or its "convergence", if it is low or high; whether they are long or short depending on the completeness of TENs.	Quantitative and qualitative combining the quantitative and qualitative tools from steps 1-4.

3.7. Conclusion

This chapter has proposed a methodological guide that can be used for the diagnosis, analysis and study of NISs in Africa. Our work and the literature show that a contextualization of the tools provided by the model is necessary. To meet this methodological challenge, we have reviewed the approaches used in agricultural innovation in Burkina Faso: the integrated approach to cash crops, the training and visit approach, the innovation platform approach and the innovation system approach. These different approaches have all been implemented in several African countries but especially in Burkina Faso. To study an NIS, we proposed and characterized six essential diagnostic steps adapted to the context of African countries, especially those sharing similar macroeconomic contexts to Burkina Faso. This methodological proposal for the diagnosis of AISs in Africa is part of a more global effort to document African experiences.

3.8. References

Akrich, M., Callon, M., Latour, B. (2002). The key to success in innovation. Part I: The art of interessement. *International Journal of Innovation Management*, 6(2), 187–206.

Belloncle, G. (1987). *Recherche, vulgarisation et développement rural en Afrique Noire.* Ministère de la Coopération, Paris.

Bjaras, G., Haglung, A., Kifkin, B. (1991). A new approach to community participation assessment. *Health Promotion International*, 6(3), 199–206.

Callon, M. (1991). Techno-economic networks and irreversibility. In *A Sociology of Monsters: Essays on Power, Technology and Domination*, Law, J. (ed.). Routledge, London.

Callon, M. and Muniesa, F. (2003). Les marchés économiques comme dispositifs collectifs de calcul. *Réseaux*, 21(122), 189–233.

Callon, M., Laredo, P., Rabeharisoa, V. (1992). The management and evaluation of technological programs and the dynamics of techno-economic networks: The case of the AFME. *Research Policy*, 21, 215–236.

Callon, M., Laredo, P., Mustar, P. (1995). *La gestion stratégique de la recherche et de la technologie*. Economica, Paris.

Casadella, V. and Benlahcen, M. (2006). De l'applicabilité du Système National d'Innovation dans les Pays Moins Avancés. *Innovations*, 2, 59–90.

Cimoli, M., Dosi, G., Stiglitz, J.E. (2009). *Industrial Policy and Development: The Political Economy of Capabilities Accumulation*. Oxford University Press.

CIRAD GRET (2002). Faciliter l'émergence et la diffusion des innovations. *Memento de l'agronome*. CIRAD, Ouagadougou.

Compaoré, E.M.F.W. (2009). Science, Technology and Development: Case Study of Agricultural Biotechnology in Burkina Faso. Institute for Science and Society, University of Nottingham.

Compaoré, E.M.F.W. (2015). The role of the national innovation systems framework facilitating in socio-economic development in Burkina Faso: Model and policy practice. PhD Thesis, University of Nottingham.

Dakouo, D. (2008). Document national de synthèse des états des lieux des résultats de recherche et des innovations du Burkina Faso. Document, MESSRS, IDR, Ouagadougou.

Dufumier, M. (1996). *Les projets de développement agricole : manuel d'expertise*. CTA-Karthala, Paris.

Edquist, C. (1997). Systems of innovation approaches – Their emergence and characteristics. In *Systems of Innovation Technologies, Institutions and Organisations*, Edquist, C. (ed.). Pinter, London.

Evanson, E.R. (2002). Recherche, vulgarisation et productivité agricole en Afrique de l'ouest. In *National Innovation Systems Overview and Country Cases. Working Paper. Center for Science, Policy, and Outcomes,* Feinson, S. (ed.). Columbia University Washington, DC, and School of Public Policy, Georgia Tech, Atlanta, GA.

Feinson, S. (2003). *National Innovation Systems, Overview and Country Cases. Knowledge Flows, Innovation and Learning in Developing Countries*. OECD, Paris.

Freeman, C. (1982). Technology infrastructure and international competiveness. Draft paper submitted to the OECD ad hoc group on Science, Technology and Competiveness.

Freeman, C. (1987). *Technology Policy and Economic Performance: Lessons from Japan*. Pinter, London.

Freeman, C. (1995). The national system of innovation in historical perspective. *Cambridge Journal of Economics*, 19(1), 5–24.

Freeman, C. and Lundvall, B. (1988). *Small Countries Facing the Technological Revolution.* London, Pinter.

Gaillard, J. (1999). *La coopération scientifique et technique avec les pays du Sud : peut-on partager la science ?*, Karthala Éditions, Paris.

Glin, C.L. (2010). Analyse des cas : platformes multi-acteurs, réalité ou nouvelle trouvaille sémantique ? In *Les plates-formes multi-acteurs dans le système national de la recherche agricole du Bénin*, Vodouhê, G.F., Lançon, J., Vodouhê, S.D. (eds). OBEPAB: Coopération française, Cotonou.

Godin, B. (2009). National innovation system : The system approach in historical perspective. *Science, Technology & Human Values*, 34(4), 476–501.

Goto. (2000). Japan's national innovation system: Current status and problems. *Oxford Review of Economic Policy*, 16(2), 103–113.

Katz, E., Riaz, J., De Meyer, B., Dovov Nichterlein, K. (2014). Approaches to strengthening agricultural innovation systems (AIS). In *Central Asia, South Caucasus and Turkey: Insights Based on Assessments of Strengths and Weaknesses of the AIS in Kyrgyzstan, Tajikistan, Uzbekistan, Azerbaijan and Turkey*. FAO, Rome.

Koutsouris, A. (2012). Facilitating agricultural innovation systems: A critical realist approach. *Studies in Agricultural Economics*, 114, 64–70.

Lundvall, B.A. (1985). *Product Innovation and User-producer Interaction.* Aalborg University Press.

Lundvall, B.A. (1992). *National Systems of Innovation: Towards a Theory of Innovation and Interactive Learning*. Pinter, London.

Lundvall, B.A. (2007). National innovation systems – Analytical concept and development tool. *Industry and Innovation*, 14(1), 95–119.

Lundvall, B.A. and Borrás, S. (1997). *The Globalising Learning Economy: Implications for Innovation Policy*. DG XII, Commission of the European Union, Brussels.

Manley, K. (2003). Frameworks for understanding interactive innovation processes. *The International Journal of Entrepreneurship and Innovation*, 4(1), 25–36.

MESSRS (1987). Symposium national de Farako-Ba : une recherche scientifique de masse au service du développement independent, autosuffisant et planifie du Burkina Faso. Summary report, CNRST, Ouagadougou.

MESSRS (1994). Forum national sur la recherche scientifique au Burkina Faso, document de base : Situation institutionnelle de la recherche scientifique et technologique au Burkina Faso. Document, MESSRS, Ouagadougou.

MESSRS (1999). Memorandum sur la situation à l'Université de Ouagadougou. Memo, UO, Ouagadougou.

MESSRS (2008). Rapport de l'état des lieux auprès des départements ministériels, des 13 regions et des partenaires stratégiques et limitrophes du Burkina Faso dans le cadre du projet. Report, FRSIT, Ouagadougou.

Miettenen, R. (2013). *Innovation, Human Capabilities, and Democracy*. Oxford University Press.

Millogo, C. (2013). Plateforme d'innovation multi-acteurs pour le transfert de technologies comme facteur d'amélioration de la performance agronomique chez les producteurs au Burkina Faso : cas des entrepreneurs agricoles de la Sissili. Thesis, Université de Ouagadougou.

Morrison, M. and Morgan, S.M. (1999). *Models as Mediators, Perspective in Natural and Social Science*. Cambridge University Press.

MRSI (2012). Politique Nationale pour la Recherche Scientifique et Technique (PNRST) 2013-2025. Report, MRSI, Ouagadougou.

Muchie, M., Gammeltoft, P., Lundvall, B.A. (2003). *Putting Africa First, the Making of Africa Innovation Systems*. Aalborg University Press.

Nauleau, G. (1987). Une opération de motorisation intermédiaire au Burkina Faso. In *Recherche, vulgarisation et développement rural en Afrique Noire*, Belloncle, G. (ed). Ministère de la Coopération, Paris.

Nelson, R.R. (1993). *National Innovation Systems: A Comparative Analysis*. Oxford University Press.

OECD (1999). *Managing National Innovation Systems*. OECD Publications, Paris.

OECD (2013). *Les systèmes d'innovation agricole. Cadre pour l'analyse du rôle des pouvoirs publics*. OECD, Paris.

Pickering, C.D. (1987). L'expérience de la Banque Mondiale en matière de vulgarisation agricole. In *Recherche, vulgarisation et développement rural en Afrique Noire*, Belloncle, G. (ed.). Ministère de la Coopération, Paris.

Russell, J. (1987). La Banque Mondiale et la vulgarisation agricole : l'introduction du système "Formation et. Visites" en Afrique Subsaharienne. In *Recherche, vulgarisation et développement rural en Afrique Noire*, Belloncle, G. (ed.). Ministère de la Coopération, Paris.

Sawadogo/Compaoré, E.M.F.W. (2016a). Transition politique et accumulation capitaliste : le cas du commerce du coton au Burkina Faso post révolutionnaire. In *Développement endogène de l'Afrique et mondialisation : une relecture de la pensée du Professeur Joseph Ki-Zerbo*, Sanon, F. and Sissao, A.J. (eds). Fondation Joseph Ki-Zerbo, PUO, Ouagadougou.

Sawadogo/Compaoré, E.M.F.W. (2016b). Monsters in agriculture: Moral controversy around genetically modified cotton in Burkina Faso. In *Writing Abnormality/Écriture de l'anormalité, A Collective Book*, N'Guessan, K.G. (ed.). Pressse Universitaire de Lomé.

Sawadogo/Compaoré, E.M.F.W. (2017). Le coton biotechnologique au Burkina Faso : trajectoire d'une innovation dans un contexte de développement, numéro spécial de technologie innovation sur l'innovation agro-écologique et de développement. *Innovations agro-écologiques et Développement*, 2, 27–37.

Sawadogo/Compaoré, E.M.F.W. and Sawadogo, N. (2018). Capacity development for agricultural innovation systems in Burkina Faso: What's new with CDAIS project? In *Innovation and Interdisciplinary Solutions for Underserved Areas*, Kebe, C.M.F., Gueye, A., Ndiaye, A. (eds). Springer International Publishing, EAI, Berlin.

Sawadogo, N. and Sawadogo/Compaoré, E.M.F.W. (2018). Theorising Africa's development problem. In *The Development of Africa: Issues, Diagnoses and Prognoses*, Olanyinka, A. and Adesina, J.O. (eds). Springer International Publishing, Berlin.

Sharif, N. (2004). Contributions from the sociology of technology to the study of innovation systems. *Knowledge, Technology, & Policy*, 17(3-4), 83–105.

Sharif, N. (2006). Emergence and development of the national innovation systems concept. *Research Policy*, 35(5), 745–766.

Sharif, N. (2009). Rhetoric of innovation policy making in Hong Kong using the innovation systems conceptual approach. *Science, Technology and Human Value*, 35(3), 408–434.

Touzard, J.M., Temple, L., Faure, G., Triomphe, B. (2015). Systems and knowledge communities in the agriculture and agrifood sector: A literature review. *Journal of Innovation Economics & Management*, 17, 117–142.

UNESCO (1999). Science for the twenty-first century: A new commitment. In *World Conference on Science*. UNESCO, Paris.

United Nations (2009). *Investment Policy Review of Burkina Faso*. United Nations, New York.

World Bank (1998). *Knowledge for Development*. World Bank, Washington, DC.

4

Adoption of Rice Technological Innovations for Technical Efficiency in Senegal

4.1. Introduction

In the majority of countries in Asia, Africa, the Americas and the Caribbean, rice is the staple food (OECD 2017). Rice is the most consumed cereal in the world and ranks third in global cereal production after wheat and maize (Seck et al. 2012). Africa is one of the world's largest importers of rice, with 40% of its needs met by imports from Asia (Boutsen and Aertsen 2013). According to World Bank forecasts, milled rice production is estimated at 509.2 million tons (FAO 2020).

In West Africa, Senegal is one of the countries where rice consumption is very high, with an annual average per capita consumption estimated at 90 kg. Indeed, since colonization, broken rice has become a staple food for Senegalese people. Annual consumption is estimated at 1,080,000 tons, of which 650,000 tons come from imports and 430,000 tons from local production (PRACAS 2014).

However, the government of Senegal has made significant investments in the sector in recent years with the National Rice Self-Sufficiency Program (*Programme National d'Autosuffisance en Riz* (PNAR)) in 2008, the Great Agricultural Offensive for Food and Abundance (*Grande Offensive Agricole pour la Nourriture et l'Abondance* (GOANA)) in 2010, and the Program for the Recovery and Acceleration of Senegalese Agriculture (*Programme de Relance et d'Accélération de la Cadence de l'agriculture Sénégalais* (RACAS)) in 2014. However, these programs have not achieved the objectives in terms of food self-sufficiency, security

Chapter written by Babacar NDIAYE and Albertine Bayompe KABOU.

and sovereignty (Ngom et al. 2016). Therefore, the opportunities offered by technological innovations vis-à-vis improved seeds, fertilizer and agricultural materials become key opportunities to improve technical efficiency in developing economies (Nkamleu 2004). The concept of technical efficiency is defined as a situation in which as much output is produced from a given set of inputs (Farrel 1957). Since then, this concept has evolved according to applications related to several fields (Maris 1997; Demeuse et al. 2005; Nuama 2010).

Some authors (Khan et al. 2017; Churchill et al. 2020; Gero 2020) explain the low productivity in most developing countries by the lack of technical efficiency. This is because most farmers in developing countries operate below their potential production capacity (Thrirtl et al. 1995; Keane et al. 2009). This is the case in Senegal, where on average, rice farmers in the Senegal River basin are not efficient (Ngom et al. 2016). These authors showed that rice farmers record a 30% gap between observed and realized production. This gap is explained by the fact that the majority of producers are not on the production and cost frontier. Thus, increasing agricultural productivity requires the introduction of technological change (Gillis et al. 1990; Nuama 2006). Using panel data on eight francophone African countries over the period 1970–2000, technological change was the main cause of the failure of total factor productivity in these countries (Nkamleu 2004).

Moreover, the adoption of new agricultural technologies is considered an alternative way out of poverty for several developing economies (Bandiera et al. 2006). Empirical impact assessment studies show that the adoption of these technologies is conditioned by observable and unobservable variables (Adesina and Zinnah 1993; Adekambi 1995; Alene and Manyong 2006; Hailu et al. 2014). Thus, several works have attempted to assess the impact of the adoption of new technologies using various estimation methods, such as the binomial regression method (Heckman et al. 1998). Indeed, accounting for heterogeneity in the adoption of new technologies is known to be an important feature of program impact evaluation methods (Heckman et al. 1998; Deb and Trivedi 2006). Nevertheless, more recent work on technological advances is proposed with machine learning models to identify farmer adoption patterns of several agricultural technologies (Lampach and Nguyen 2019).

Our chapter differs from work of Duflo et al. (2011) and Duval (2017), who have defined treatment, that is, participation in the project or program, in a binary way (yes or no). Considering levels of treatment as the adoption of new rice technologies, we disaggregated the different possibilities or treatments that the programs offer using multinomial logistic regression. These treatments are fertilizer, improved seed and motorized equipment. Thus, treatment level T1 is for the farmers who adopted only fertilizer, treatment level T2 is for all farmers who adopted fertilizer and improved seed simultaneously, and the final treatment level T3 is for the farmers who combined fertilizer, improved seed and motorized equipment.

This chapter is organized as follows. The first section presents the theoretical framework of the estimation methods. The second section presents the model specifications for detecting the impact of each treatment level on the technical efficiency of rice farmers. The conclusion presents discussions and research perspectives.

4.2. Theoretical framework

4.2.1. *Theoretical approach to the determinants of adoption*

The literature on the determinants of adoption highlights several characteristics that influence adoption status. Adoption refers to several concepts in the literature (Ahmed et al. 2017; Hochscheid and Halin 2018). Ahmed et al. argue that adoption is a decision made most often by a company's management to signal to workers the introduction of a new technology into their work. It is considered a process of integrating an innovation (Rogers 2003), which is a combination between the concepts of implementation and diffusion (Succar and Kassem 2015). Thus, the adoption of new technologies is a form of "process innovation" (management and production methods) as opposed to "product innovation" (creation of new products with certain characteristics that are different from previous products) (Lefebvre et al. 1990).

The role of women has been demonstrated in the adoption process (Adesina and Baidu-Forson 1995). These authors show that in West Africa, women are in the best position to appreciate the quality of products derived from new technologies. In Malawi, Chirwa (2005) finds that female-headed households are the least affected by the adoption of new technologies. This is because female-headed households are more likely to be poor and have limited resources, which reduces their ability to adopt new technologies.

Regarding age, Alene and Manyong (2005) show that relatively old farmers are less equipped to use some new technologies efficiently. On the other hand, the emergence of new tools has made it easier for this category of people to adopt new technologies (Adesina and Baidu-Forson 1995).

Several financial factors can also motivate the adoption of new technologies. Research has shown that in developing countries, the non-existence or imperfection of the insurance market is a real obstacle to the adoption of new technologies (Udry 2010). In addition, difficulties in accessing credit can also be a handicap in the adoption of new technologies, especially during non-harvest periods (Duflo et al.

2011). As a result, farmers often face liquidity problems. To address these constraints, the availability of and access to credit will enhance the use of certain inputs (Zeller 1998; Alene and Manyong 2006; Hailu et al. 2014).

During the green revolution in China (1960–1990), coinciding with a political and industrial will supported by scientific and technical progress in the agricultural field, the adoption of new agricultural technologies led to a significant increase in agricultural productivity (World Bank 2008). In the same sense, many studies show that the adoption of improved technologies increases agricultural productivity (Mendola 2007; Seck et al. 2012).

4.3. Empirical approach

Throughout the economic literature, several works have addressed farmers' choice to adopt new technologies. Adékambi, with particular emphasis on investment in children's schooling and health, assesses the socioeconomic impact of improved rice varieties in Benin using a simple comparison between adopters and non-adopters. Mayen et al. (2010) used the propensity score that corresponds to production frontier analysis to compare productivity between conventional and dairy farms in the United States. Using the same approach, Martey et al. (2015) evaluate the impact of mentoring project participation on farm efficiency. Villano et al. (1995) use propensity score matching in production frontier analysis to compare technical inefficiency between adopters and non-adopters of certified seed in the Philippines.

Technical efficiency argues that the producer has maximum potential output, given the factors of production (inputs) available to him. Methods for calculating technical efficiency can be grouped into two broad families: parametric and non-parametric methods (Ngom et al. 2016). However, in the agricultural sector, parametric production frontier estimation is more appropriate (Aigner et al. 1977; Meeusen 1977; Ngom et al. 2016). Unlike the agricultural process envelopment method, this method includes the determinants of the technical inefficiency of agricultural households (Lampach and Morawetz 2016).

To assess the impact of the adoption of new technologies on efficiency, it is important to refer to the literature on adoption, which refers to the decision to implement new technical proposals that depend on the socioeconomic characteristics of potential adopters (Feder and Umali 1993; Rogers 2003). Beyond this approach, technical proposals also depend on the compatibility of the characteristics of the innovations with the institutional (norms, rules, values), technological (existing technical systems, know-how, risk) and economic (accessibility of the necessary production factors) environment of the potential adopters but also on their perception of the characteristics of the innovations proposed to them (Rogers 2003). Thus, the allocation of the treatment is distributed in a non-random way in the population, and the motive for adoption is determined either by observable variables or by unobservable variables. The challenge is to identify a counterfactual, that is, an appropriate comparison group similar to the treated group.

Although the propensity score matching approach is the most common method in public policy evaluation, it leads to inefficient and inaccurate biased estimates (Lampach and Morawetz 2016). This is because the propensity score matching method is a predicted probability of treatment assignment based on observed characteristics. Thus, the propensity score matching method eliminates the bias induced by observable characteristics (Mulugeta and Hundie 2012). These authors use this method to assess the potential impact of agricultural technology adoption on household food consumption status.

However, several limitations are noted on simple regressions because the bias related to self-selection of collected observation data is not eliminated (Rubin 1974; Heckman and Vytlacil 2005). Thus, to consider this bias, Mulugeta and Hundie (2012) developed theories through experimental and quasi-experimental methods to address the self-selection problem. The limitation of this method is that it only controls for observable variables that are part of the variables driving self-selection in the adoption process. As a result, only the bias related to the observable variables is removed (Rosenbaum 2002; Lee 2005).

Producers anticipate their returns in the case of new technology adoption (Imbens and Angrist 1994). The adoption of new technologies depends on unobserved characteristics and macroeconomic factors affecting the incremental productivity of the technology compared to traditional technologies (Suri 2011). To eliminate biases related to observable or unobservable characteristics and address the endogeneity problem of the treatment variable, several authors use instrumental variables (Imbens and Angrist 1994; Abadie 2001; Dontsop Nguezet et al. 2012). The role of the instrumental variable is to introduce exogenous variation into the treatment variable by offering a causal interpretation (Abadie 2001). This method takes into account the problem of endogeneity related to selection but is only applied in the binary case.

4.4. Econometric model and data analysis

4.4.1. *Specification of the production boundary method*

The data are from the Agricultural Policy Support Project (PAPA) survey for irrigated rice and the Directorate of Agricultural Analysis, Forecasting and Statistics

(DAPSA) for rainfed rice in 2017. The sample is composed of 1,200 households, of which 526 choose irrigated rice and 674 choose rainfed rice.

The estimation of the production frontier is analyzed using the simplified log transcendental function. This method has, on the one hand, the advantage of being flexible and does not impose any restrictive assumptions on the constant and elasticity of the production function (Donkoh et al. 2013). On the other hand, it allows us to see the combination between two factors of production on the level of production. Therefore, the function can be written as follows:

$$lnY_i = \beta_0 + \sum_{j-1}^{5} \beta_j lnX_j + \sum_{j-1}^{5} \sum_{l-1}^{5} \beta_m lnX_j lnX_l + (v_i - u_i) \quad [4.1]$$

with:

Y_i is the total harvested production in kilograms; X_{1i} is the total area in acres cultivated; X_{2i} is the the total amount of seed used in kilograms; X_{3i} is the total amount of fertilizer (npk and urea) used in kilograms; X_{4i} is the amount of labor expressed in person/day; X_{5i} is the quantity of equipment used; and β is the vector of parameters to be estimated.

To estimate the impact of new rice technologies on technical efficiency, the use of the *translog* functional form seems most appropriate. This choice is justified by the most appropriate specification between the Cobb–Douglas function and the *translog* function. Thus, the rejection of the null hypothesis is induced by the fact that $\lambda = 69.19$ is higher than the critical value.

The stochastic production function model takes the following form:

$$\begin{aligned} lnY_i = {} & \beta_1 lnarea_i + \beta_2 lnseed_i + \beta_3 lnfert_i + \beta_4 lnlabor_i + \beta_6 lnequip_i + \\ & \beta_7 lnarea_i{}^2 + \beta_8 lnseed^2 + \beta_9 lnfert_i{}^2 + \beta_{10} lnlabor_i{}^2 + \beta_{11} lnequip_i{}^2 + \\ & \beta_{12} lnarea_i * lnseed_i + \beta_{13} lnarea_i * lnfert_{i0} + \beta_{14} lnarea_i * lnlabor_i + \\ & \beta_{15} lnarea_i * lnequip_i + \beta_{16} lnseed_i * lnfert_i + \beta_{17} lnseed_i * lnlabor_i + \\ & \beta_{18} lnseed_i * ln\acute{e}quip_i + \beta_{19} lnfert_i * lnlabor_i + \beta_{20} lnfert_i * lnequip_i + \\ & \beta_{21} lnlabor_i * lnequip_i + \epsilon_i \end{aligned} \quad [4.2]$$

lnY_i is the log of household rice production in tons i. Inputs (area, seed, npk and urea fertilizer, labor and farm equipment) are expressed in log. The parameters β_1 to β_{21} are to be estimated, and ϵ_i denotes the error term, including a noise term v_i and a technical inefficiency u_i.

With the multinomial choice model, each individual i makes a choice on a set of two or three treatments. Let EV_{ij} correspond to the indirect utility obtained by selecting a treatment j ranging from $j = 0, 1, 2$ and 3.

$$EV_{ij} = z'_i\alpha_j + \sum_{k=1}^{j} \delta_{jk}l_{ik} + \eta_{ij} \quad [4.3]$$

where z_i corresponds to exogenous covariates for which the associated parameters α_j and η_{ij} are independently and identically distributed error terms, EV_{ij} includes a latent factor l_{ij} that incorporates unobserved characteristics common to the treatment choice and the individual's outcome i. l_{ij} is assumed to be independent of η_{ij}. We consider that $j = 1$ denotes the control group with $EV_{i0} = 0$.

Let d_j be the variables that represent the observed treatment choice with $d_i = (d_{i1}, d_{i2}, d_{ij})$.

The probability of treatment can be written as follows:

$$p_r(d_i/z_i, l_i) = g(z_i'\alpha_1 + \sum_{k=1}^{j} \delta_{1k}l_{ik}, z_i'\alpha_2 + \sum_{k=1}^{j} \delta_{2k}l_{ik}, \ldots, z_i'\alpha_j + \sum_{k=1}^{j} \delta_{jk}l_{ik}) \quad [4.4]$$

where g is an appropriate multinomial probability distribution.

The expected outcome equation for the individual i, $i = 1, 2, \ldots, N$ is formulated as follows:

$$E(y_i) = \mu(z_i'\beta + \sum_{j=1}^{j} \gamma_j d_{ij} + \sum_{j=1}^{j} \lambda_j l_{ij}) \quad [4.5]$$

$E(y_i)$ is a function of each of the latent factors l_{ij}, that is, the outcome is affected by unobservable characteristics that also affect selection for treatment.

4.4.2. *Analysis of results*

The similarity analysis of the two groups shows that there is a difference between the average of the treated and untreated[1]. At treatment level T1, there are 273 households in the treated group and 927 households in the untreated group. For treatment level T2, there are 357 adopter households and 843 non-adopter households. For treatment level T3, the number of adopters is 112 households and 1088 non-adopters.

1 See Table 4.3 in Appendix 1 (section 4.6).

Variables	Coef.	Std. Err.	P>|z|
	T1 (fertilizer)		
Contract	1.075	0.782	0.169
Credit	1.914	0.815	0.019 **
M_OP	1.026	0.291	0.000 ***
Insur.	0.788	0.668	0.238
Female	0.333	0.314	0.289
Single	0.051	0.803	0.949
Age_cm	–0.001	0.007	0.907
Household_size	0.011	0.018	0.530
Tec_access	4.455	1.022	0.000 ***
Cons	–1.036	0.456	0.023 **
	T2 (fertilizer and improved seeds)		
Contract	0.440	0.816	0.589
Credit	3.912	0.781	0.000 ***
M_OP	1.912	0.300	0.000 ***
Insur.	2.111	0.641	0.001 ***
Female	–0.699	0.430	0.104
Single	–1.455	0.803	0.315
Age_cm	–0.015	0.009	0.092 *
Household_size	–0.023	0.021	0.312
Tec_access	6.803	1.018	0.000 ***
Cons	–0.950	0.532	0.074 *
	T3 (fertilizer, improved seeds and motorized equipment)		
Contract	1.859	0.790	0.016 **
Credit	3.479	0.809	0.000 ***
M_OP	2.223	0.332	0.000 ***
Insur.	2.777	0.648	0.000 ***
Female	-0.948	0.603	0.116
Single	-0.358	1.318	0.786
Age_cm	-0.021	0.11	0.056 *
Household_size	0.016	0.26	0.547
Tech_access	4.412	1.043	0.000 ***
Cons	–2.815	0.180	0.000 ***

NOTE.– ***, ** and * significance at 1%, 5% and 10%.

Table 4.1. *Determinants of technology adoption by treatment level. Source: Authors, using AHS and DAPSA survey data (2017)*

Table 4.1 presents the results of the probit model for estimating the determinants of adoption of the new rice techniques at three levels of treatment.

Number of observations = 1,124

Wald χ^2 (27) = 1787.59

Prob > χ^2 = 0.0000

The results show that the model is globally significant at the 1% level. The adoption of treatment level T1 (fertilizer) is determined by *credit, membership in a farmer organization* (m_OP) and *access to new technologies* (tech_access). The determinants of treatment T2 (fertilizer and improved seed) are *credit, membership in a farmer organization* (m_OP), *insurance, access to technology* and *age of the household head* (age_cm). The adoption of treatment T3 (fertilizer, improved seeds and motorized equipment) is strongly influenced by the variables *contract, credit, membership in a farmer organization, insurance, access to new technologies* and *the age of the household head.*

The coefficient of the *credit* variable is positive and significant for treatment T1 at the 5% threshold and for treatments T2 and T3 at the 1% threshold. The fact that rice farmers have credit is a positive factor in the adoption of the three levels of treatment. This result is explained by the fact that holding credit favors the adoption of new technologies. This result corroborates that of Hailu et al. (2014), for whom access to credit enhances the use of certain inputs.

The variables *membership in a farmer organization* (m_OP), *access to technology* (tech_access) and *insurance* have a positive and significant impact on the adoption of the three levels of treatment at the 1% threshold. This result shows that in Senegal, farmer organizations play an important role in encouraging the adoption of new technologies. Baumüller (2012) argues that awareness within farmers' organizations is important in decisions to adopt a new technology in agriculture. In addition, the uncertainties associated with the adoption of a new technology mean that buying insurance allows the producer to protect himself from the risks involved. According to Udry (2010), farmers without insurance protection or with imperfect insurance will be more reluctant to adopt.

In contrast, the *age of the household head* (age_cm) variable negatively impacts the adoption of all three levels of treatments T1, T2 and T3. This result indicates that older people have less incentive to adopt new technologies. Chirwa (2005) has shown that relatively old farmers are the most reluctant to adopt new technologies.

Number of observations = 1,124

Wald χ^2 (27) = 1,787.59

Prob > χ^2 = 0.0000

Variables	Coef.	Std. Err.	P > |z|
T1 (fertilizer)	0.015	0.016	0.370
T2 (fertilizer and improved seeds)	0.097	0.015	0.000 ***
T3 (fertilizer, sem_improved and equip_moto)	0.121	0.022	0.000 ***
Nbr_parcel	0.004	0.002	0.153
Actv_principal	–0.004	0.019	0.807
Female	–0.327	0.017	0.000 ***
Single	–0.348	0.044	0.000 ***
Married_poly	–0.372	0.015	0.000 ***
Literate	0.273	0.203	0.000 ***
No_for_educ	0.217	0.016	0.000 ***
Age_cm	–0.003	0.000	0.000 ***
Culture_season	–0.521	0.019	0.000 ***
Household_size	–0.003	0.001	0.000 ***
Cons	–3.634	0.056	0.000 ***
/lnalpha	3.903	0.046	0.000 ***
/lambda_T1	0.008	0.013	0.498
/lambda_T2	–0.011	0.010	0.287
/lambda_T3	–0.014	0.011	0.192
Alpha	49.575	2.296	

NOTE.– ***, ** and * significance at 1%, 5% and 10%.

Table 4.2. *Impact of new technology adoption on technical efficiency. Source: Authors, using AHS and DAPSA survey data (2017)*

Table 4.2 presents the impact of the adoption of new rice technologies on technical efficiency. The model is globally significant at the 1% level for all three treatment levels. The first treatment level T1 has no impact on the farmer's technical efficiency.

The adoption of the T2 treatment level impacts the technical efficiency of Senegalese rice farmers by 9.7% at the 1% threshold. Thus, the combination of fertilizer and improved seed increases the level of technical efficiency by 9.7%. This result is similar to that of Hailu et al. (2014), who show that the adoption of chemical fertilizer and high-yielding seed increases farmers' income in northern Ethiopia.

Adoption of the T3 treatment increases the level of technical efficiency by 12.1%. This result shows that farmers who use a combination of fertilizer and improved seed have lower technical efficiency than those who use a combination of fertilizer, improved seed and motorized equipment. Thus, the difference in technical efficiency between those using the T2 treatment and the T3 treatment is 2.4 percentage points. These results are identical to those of Mendola (2007) and Seck et al. (2012), who argue that the adoption of new technologies allows for increased productivity. Unlike these results, our methodological approach considers a multinomial logistic regression.

Furthermore, at the 1% threshold, the *female* variable negatively impacts technical efficiency by 32.7%. This result shows that the yields of rice farms managed by women are relatively less efficient than those managed by men. The *single and married_polygamous* (married_polyg) variables negatively impact technical efficiency by 34.8% and 37.2%, respectively. The *age of the head of household* variable (age_cm) reduces technical efficiency by 0.3%. This result corroborates that of Alene and Manyong (2006), who show that marital status has an impact on the way certain new technologies are adopted. *Household size* (household_size) negatively impacts technical efficiency by 0.3%. This result suggests that the production efficiency of large families is relatively low because much of the labor force is not used on the rice farm. The process of mechanization in terms of motorized equipment therefore impacts the number of people in the household.

Nevertheless, some variables, such as *literacy* and *no formal education* (no_form_education), positively impact technical efficiency by 27.3% and 21.7%, respectively, at the 1% threshold.

4.5. Conclusion

Senegal is one of the countries in West Africa with the highest rice consumption. Its national production covers between 20 and 30% of the country's rice demand (Ministère de l'Agriculture 2014). This production deficit has been the subject of several interventions by the state through programs without achieving the goal of self-sufficiency. However, rice production varies from year to year under the influence of various factors, such as rainfall conditions, soil conditions, availability of inputs, seed quality and intensity of mechanization. It is in this context that the improvement of inputs through the opportunities offered by new rice technologies becomes appropriate.

This chapter highlights the determinants of the adoption of new technologies (fertilizer, improved seeds and motorized equipment). Among these determinants, the variables *credit, membership in a farmer organization* and *access to technologies* positively affect the adoption status of the three levels of treatments, while the variable *insurance* influences only the adoption of treatments T2 and T3. The variable *age of the household head* negatively affects adoption status, as relatively older people are more reluctant to adopt innovations than younger people.

Our results also reveal that technical efficiency increases with the level of treatment. Indeed, when the farmer adopts treatment T1, the impact on technical efficiency is not significant. On the other hand, when the farmer adopts treatment T2 (fertilizer and seed), the impact on technical efficiency is 9.7% at the 1% threshold. This result shows that the farmer's use of a combination of fertilizer and improved seed has a greater impact on technical efficiency than the sole use of fertilizer (T1). We also find that the adoption of treatment T3 (fertilizer, improved seed and motorized equipment) impacts the farmer's technical efficiency by 12.1%, an increase of 2.4 percentage points compared to treatment T2. According to these results, Senegalese rice farmers who use fertilizer, improved seed and motorized equipment simultaneously are more efficient than those who use fertilizer and improved seed alone.

In terms of policy implications, this chapter shows how important it is to invest in new agricultural technologies to combat poverty and food insecurity. Thus, establishing effective channels for the distribution of these new rice technologies and training users are essential in rural empowerment strategies. This allows the highest level of processing to be adopted, especially when other innovations are integrated into the agricultural production process.

The main limitation of this work is that there are several types of improved seeds whose characteristics are not identifiable from the database. Therefore, one of the perspectives of this research is to consider mining data on the types of improved seeds to differentiate their level of effectiveness. In terms of estimation method, impact assessment methods such as the instrumental method are the most appropriate, but they are only applied in the binary case, whereas our approach is in the multiple case, hence the use of the Deb and Trivedi (2006) model.

4.6. Appendix 1

Variables	Definitions	T1 (fertilizer)		T2 (fertilizer and improved seeds)		T3 (fertilizers, improved seeds and motorized equipment)	
		Adopt	Non-adopt	Adopt	Non-adopt	Adopt	Non-adopt
Nbr_parcel	Number of plots cultivated by rice farmers	2.02 (272)	2.16 (924)	1.67** (356)	2.33** (840)	3.20*** (112)	2.02*** (1,088)
Contract	Benefited from an agricultural contract 1 = yes; 0 = no	0.03 (927)	0.04 (273)	0.05 (357)	0.04 (843)	0.18*** (112)	0.03*** (1,088)
Credit	Benefited from an agricultural credit 1 = yes; 0 = no	0.04*** (273)	0.98*** (927)	0.16*** (357)	0.05*** (843)	0.18*** (112)	0.07*** (1,088)
M_OP	1 = number of a farmer organization; 0 = no	0.21** (273)	0.28** (927)	0.44*** (357)	0.19*** (843)	0.51*** (112)	0.24*** (1,088)
Act_prin	1 = agriculture; 0 = no	0.95 (273)	0.94 (927)	0.98*** (357)	0.93*** (843)	0.99** (112)	0.94** (1,088)
Female	1 = woman; 0 = no	0.12** (273)	0.07** (927)	0.06** (357)	0.10** (843)	0.04* (112)	0.09* (1,088)
Single	1 = single; 0 = no	0.14 (273)	0.01 (927)	0.002** (357)	0.01** (843)	0.008 (112)	0.01 (1,088)
Married_polyg	1 = polygamous married; 0 = not married	0.30 (273)	0.24 (927)	0.22*** (357)	0.31*** (843)	0.33 (112)	0.27 (1,088)
Literate	1 = literacy; 0 = otherwise	0.07 (273)	0.08 (927)	0.13*** (357)	0.06*** (843)	0.24*** (112)	0.06*** (1,088)
Age_cm	Age of head of household	53.84 (273)	52.75 (927)	51.66** (357)	53.57** (843)	51.84 (112)	53.12 (1,088)
Culture_type	1 = irrigation; 0 = no	0.42 (273)	0.44 (927)	0.74*** (357)	0.30*** (843)	0.98*** (112)	0.38*** (1,088)
Household_size	The number of people in the household	9.59 (273)	9.23 (927)	8.59*** (357)	9.62*** (843)	10.41** (112)	9.20** (1,088)

NOTE.– 1%, 5% and 10% significance level for ***, ** and *, respectively.

Table 4.3. *Socioeconomic characteristics of adopter groups by treatment level. Source: Authors using AHS and DAPSA survey data (2017)*

4.7. Appendix 2: empirical model of technical efficiency

The stochastic frontier model for producer i, with i ∈ {1, 2, 3, 4, ..., n} is written as

$$lny_i = x_i\beta + v_i - u_i \tag{4.6}$$

where y_i represents the output of farmer i, x_i denotes a vector of inputs, that is, the vector of associated coefficients to be estimated is the error term that captures stochastic effects that are not under the control of farmer i, and u_i represents the random variable, positive or zero, reflecting technical inefficiency in terms of i's output. The latter two variables are independent and distributed according to a truncated normal distribution at zero with mean δ_i and variance σ_u^2 (N (ui, σ_u^2)). The non-negativity condition on i ensures that the observations cannot lie above the production frontier.

The agricultural production of the rice farmer i can be given by the following equation:

$$ln\hat{y}_i = x_i\beta + v_i - u_i \tag{4.7}$$

This equation represents the inefficiency-free exit from the frontier:

$$ln\ddot{y}_i = x_i\beta + v_i \tag{4.8}$$

The ratio between observed production ($ln\hat{y}_i$) and estimated production ($ln\ddot{y}_i$) on the production frontier of the perfectly efficient rice farmer using the same input vector x_i gives the estimated technical efficiency. Thus, the efficiency level of a rice farmer is obtained as follows:

$$ln\mathrm{ET}_i = \frac{ln\hat{y}_i}{ln\ddot{y}_i} = -u_i \tag{4.9}$$

The efficiency level has a value between zero and one. The objective is to measure the maximum output of the rice farmer as a proportion of the output corresponding to the optimal production frontier with the same inputs. The parameters of the frontier model are estimated in equation [4.6] by maximum likelihood estimation. The trans-log production function specification (translog) is used to estimate the likelihood function.

4.8. References

Abadie, A. (2001). Implementing matching estimators for average treatment effects in Stata. *Stata Journal*, 1(1), 1–18.

Adekambi, S.A. (2005). *Impact de l'adoption des variétés améliorées de riz sur la scolarisation et la santé des enfants au Bénin : cas du département des Collines*. FSA, UAC, Benin.

Adesina, A.A. and Baidu-Forson, J. (1995). Farmers' perceptions and adoption of new agricultural technology: Evidence from analysis in Burkina Faso and Guinea, West Africa. *Agricultural Economics*, 13(1), 1–9.

Adesina, A.A. and Zinnah, M.M. (1993). Caractéristiques technologiques, perceptions des agriculteurs et décisions d'adoption : une application du modèle Tobit en Sierra Leone. *Économie agricole*, 9(4), 297–311.

Ahmed, A., Paul Kawalek, J., Kassem, M. (2017). A conceptual model for investigating BIM adoption by organisations. In *Lean and Computing in Construction Congress (LC3): Volume 1: Proceedings of the Joint Conference on Computing in Construction (JC3)*, 533–540. Heraklion.

Aigner, D., Lovell, C.K., Schmidt, P. (1977). Formulation and estimation of stochastic frontier production function models. *Journal of Econometrics*, 6(1), 21–37.

Alene, A.D. and Manyong, V.M. (2006). Farmer-to-farmer technology diffusion and yield variation among adopters: The case of improved cowpea in northern Nigeria. *Agricultural Economics*, 35, 203–211.

Bandiera, O., Rasul, I., Andiera, O., Rasul, I. (2006). Réseaux sociaux et adoption de la technologie dans le nord du Mozambique. *Le journal économique*, 116(514), 869–902.

Barath, L. and Ferto, I. (2017). Productivity and convergence in European agriculture. *Journal of Agricultural Economics*, 68(1), 228–248.

Battese, G.E. and Coelli, T.J. (1995). A model for technical inefficiency effects in a stochastic frontier production function for panel data. *Empirical Economics*, 20(2), 325–332.

Baumüller, H. (2012). Facilitating agricultural technology adoption among the poor: The role of service delivery through mobile phones. ZEF Working Paper Series 93 [Online]. Available at: https://ssrn.com/abstract=2237987 or http://dx.doi.org/10.2139/ssrn.2237987.

Boutsen, S. and Aertsen, J. (2013). Peut-on nourrir l'Afrique de l'Ouest avec du riz ? MO Papers.

Chirwa, E.W. (2005). Adoption of fertiliser and hybrid seeds by smallholder maize farmers in Southern Malawi. *Development Southern Africa*, 22(1), 1–12.

Churchill, A., Sefa Inekwe, J., Ivanovski, K. (2020). Convergence of R&D intensity in OECD countries: Evidence since 1870. *Empirical Economics*, 59(1), 295–306.

Croppenstedt, A., Demeke, M., Meschi, M.M. (2003). Technology adoption in the presence of constraints: The case of fertilizer demand in Ethiopia. *Review of Development Economics*, 7(1), 58–70.

Darku, A.B., Malla, S., Tran, K.C. (2016). Sources and measurement of agricultural productivity and efficiency in Canadian provinces: Crops and livestock: Sources and measurement of agricultural productivity and efficiency. *Canadian Journal of Agricultural Economics*, 64(1), 49–70.

Deb, P. and Trivedi, K. (2006). Specification and simulated likelihood estimation of a non-normal treatment-outcome model with selection: Application to health care utilization. *The Econometrics Journal*, 9(2), 307–331.

Demeuse, M., Matoul, A., Schillings, P., Denooz, R. (2005). De quelle efficacité parle-t-on ? In *Vers une école juste et efficace. 26 contributions sur les systèmes d'enseignement et de formation*, Demeuse, M., Baye, A., Straeten, M.H., Nicaise, J., Matoul, A. (eds). De Boeck Université, Brussels.

Donkoh, S.A., Ayambila, S., Abdulai, S. (2013). Technical efficiency of rice production at the Tono irrigation scheme in northern Ghana. *American Journal of Experimental Agriculture*, 3(1), 25–42.

Dontsop Nguezet, P.M., Okoruwa, V., Adenagan, K.O. (2012). Productivity impact differential of improved rice technology adoption among rice farming households in Nigeria. *Journal of Crop Improvement*, 26(1), 1–21.

Duflo, E., Kremer, M., Robinson, J. (2011). Nudging farmers to use fertilizer: Theory and experimental evidence from Kenya. *The American Economic Review*, 101, 2350–2390.

Duval, L. (2017). Performance export des entreprises vitivinicoles françaises. Influence des stratégies et des politiques à l'exportation. *Économie rurale*, 6(362), 65–89.

FAO (2020). *Programme mondial du recensement de l'agriculture 2020. Vol. 2. Directives opérationnelles*. Food & Agriculture Organisation, Geneva.

Farrel, J. (1957). The measurement of productive efficiency. *Journal of the Royal Statistical Society, Series A*, 2, 252.

Feder, G. and Umali, D.L. (1993). The adoption of agricultural innovations: A review. *Technological Forecasting and Social Change*, 43(3–4), 215–239.

Gero, A.A. (2020). Innovation technologique, efficacité technique et convergence de la productivité agricole dans l'UEMOA. *Ann. Univ. Lomé, série Sc. Eco. et Gest.*, XVI, 55–67.

Gillis M., Dwight H.P., Roemer M., Snodgrass D.R. (1990). *Économie du développement*. De Boeck, Brussels.

Hailu, B.K., Abrha, B.K., Weldegiorgis, K.A. (2014). Adoption and impact of agricultural technologies on farm income: Evidence from Southern Tigray, Northern Ethiopia. *International Journal of Food and Agricultural Economics (IJFAEC)*, 2, 91–106.

Heckman, J.J. and Vytacil, E. (2005). Structural equations, treatment effects, and econometric policy evaluation. *Econometrica*, 73(3), 669–738.

Heckman, J.J., Ichimura, H., Todd, P. (1998). Matching as an econometric, evaluation estimator. *Review of Economic Studies*, 65, 261–294.

Hochscheid, É. and Halin, G. (2018). L'adoption du BIM dans les agences d'architecture en France. In *SHS Web of Conferences*, 47, 01009. EDP Sciences, Paris.

Imbens, G. and Angrist, J.D. (1994). Identification and estimation of local average. *The Econometric Society*, 62(2), 467–475.

Keane, J., Page, S., Kergna, A., Kennan, J. (2009). *Climate Change and Developing Country Agriculture: An Overview of Expected Impacts, Adaptation and Mitigation Challenges, and Funding Requirements*. ICTSD–IPC Platform on Climate Change, Agriculture and Trade, International Centre for Trade and Sustainable Development.

Khan, F., Salim, R., Bloch, H., Islam, N. (2017). The public R&D and productivity growth in Australia's broadacre agriculture: Is there a link? *Australian Journal of Agricultural and Resource Economics*, 61(2), 285–303.

Lampach, N. and Morawetz, U.B. (2016). Credibility of propensity score matching estimates. An example from fair trade certification of coffee producers. *Applied Economics*, 48(44), 4227–4237.

Lampach, N., To-The, N., Nguyen-Anh, T. (2019). Technical efficiency and the adoption of multiple agricultural technologies in the mountainous areas of Northern Vietnam. *SSRN Electronic Journal*. doi: 10.2139/ssrn.3421434.

Lee, J. (2005). Effects of leadership and leader-member exchange on commitment. *Leadership & Organization Development Journal*, 26(8), 655–672. doi: 10.1108/01437730510633728.

Lefebvre, E., Lefebvre, L.A., Colin, D. (1990). Facteurs d'adoption des nouvelles technologies de production dans les PME manufacturières. *Revue internationale*, 3(2), 215–229.

Maris, B. (1997). Qu'est-ce que l'efficacité ? In *Bernard Charbonneau : une vie entière à dénoncer la grande imposture*, Prades, J. (ed.). Érès, Toulouse. doi: 10.3917/eres.prade.1997.01.0099.

Martey, E., Dogbe, W., Etwire, P.M., Wiredu, A.N. (2015). An impact of farmer mentorship project on farm efficiency and income in rural Ghana. *Journal of Agricultural Science*, 7(10), 79.

Mayen, C.D., Balagtas, J.V., Alexander, C.E. (2010). Technology adoption and technical efficiency: Organic and conventional dairy farms in the United States. *American Journal of Agricultural Economics*, 92(1), 181–195.

Meeusen, W. and van den Broeck, J. (1977). Efficiency estimation from Cobb-Douglas production functions with composed error. *International Economic Review*, 18(2), 435–444.

Mendola, M. (2007). Adoption des technologies agricoles et réduction de la pauvreté : une analyse d'appariement du score de propension pour le Bangladesh rural. *Politique alimentaire*, 32(3), 372–393.

Ministère de l'Agriculture et de l'Equipement Rural (2015). Document principal : Programme d'Accélération de la Cadence de l'Agriculture Sénégalaise (PRACAS). Document, Volet agricole du Plan Sénégal Emergent (PSE).

Mulugeta, T. and Hundie, B. (2012). Impacts of adoption of improved wheat technologies on households' food consumption in southeastern Ethiopia. In *International Association of Agricultural Economists Conference*, August 18–24. Foz do Iguacu.

Ngom, C.A.B., Sarr, F., Fall, A.A. (2016). Mesure de l'efficacité technique des riziculteurs du bassin du fleuve Sénégal. *Économie rurale. Agricultures, alimentations, territoires*, 355, 91–105.

Nkamleu, G.B. (2004). Productivity growth, technical progress and efficiency change in African agriculture. *African Development Review*, 16(1), 203–222.

Nuama, E. (2006). Mesure de l'efficacité technique des agricultrices de cultures vivrières en Côte-d'Ivoire. *Économie rurale. Agricultures, alimentations, territoires*, 296, 39–53.

Nuama, E. (2010). The technical efficiency of rice producers in Côte d'Ivoire: Extension services in question. *Économie rurale*, 316(2), 36–47.

OECD (2016). OECD-FAO agricultural outlook 2017–2026. OECD Publishing.

Rogers, E.M. (2003) *Diffusion of Innovation*, 5th edition. Free Press, New York.

Rosenbaum, P.R. (1983). The central role of the propensity score in observational studies for causal effects. *Biometrika*, 70(1), 41–55.

Rosenbaum, P.R. (ed.) (2002). Overt bias in observational studies. In *Observational Studies*. Springer, New York.

Roussy, C., Ridier, A., Chaib, K. (2015). Adoption d'innovations par les agriculteurs : rôle des perceptions et des préférences. INRA, France.

Rubin, D.B. (1974). Estimating causal effects of treatments in randomized and nonrandomized studies. *Journal of Educational Psychology*, 66(5), 688–701.

Savadogo, K., Reardon, T., Pietola, K. (1998). Adoption of improved land use technologies to increase food security in Burkina Faso: Relating animal traction, productivity, and non-farm income. *Agricultural Systems*, 58(3), 441–464.

Seck, P.A., Diagne, A., Mohanty, S., Wopereis, M.C. (2012). Des cultures qui nourrissent le monde 7 : le riz. *Sécurité alimentaire*, 4(1), 7–24.

Succar, B. and Kassem, M. (2015). Macro-BIM adoption: Conceptual structures. *Automation in Construction*, 57, 64–79.

Suri, T. (2011). Selection and comparative advantage in technology adoption. *Econometrica*, 79(1), 159–209.

Thirtl, C., Hadley, D., Townsend, R.F. (1995). Policy-induced innovation in Sub-Saharan African agriculture: A multilateral Malmquist productivity index approach. *Development Policy Review*, 13(4), 323–342.

Toma, P., Miglietta, P.P., Zurlini, G., Valente, D., Petrosillo, I. (2017) A nonparametric bootstrap-data envelopment analysis approach for environmental policy planning and management of agricultural efficiency in EU countries. *Ecological Indicators*, 83, 132–143.

Udry, C. (2010). L'économie de l'agriculture en Afrique : notes pour un programme de recherche. *Journal africain d'économie agricole et des ressources*, 5, 284–299.

Villano, R., Bravo-Ureta, B., Solis, D., Fleming, E. (2015). Rice technologies and productivity in the Philippines: Disentangling technology from managerial gaps. *Journal of Agricultural Economics*, 66(1), 129–154.

Volet agricole du plan Sénégal émergent (2014). *Programme d'Accélération de la Cadence de l'Agriculture Sénégalaise PRACAS.* Volet agricole du plan Sénégal émergent.

World Bank (2008). World development report 2008: Agriculture for development. Report, World Bank.

Zeller, M., Diagne, A., Mataya, C. (1998). Market access by smallholder farmers in Malawi: Implications for technology adoption, agricultural productivity and crop income. *Agricultural Economics*, 19(1–2), 219–229.

5

Characterization of the Sectoral Cocoa Innovation System in Cameroon

5.1. Introduction

Introduced to Africa by the Spanish in the 16th century, Cameroonian cocoa production is one of the oldest in Africa (Alary 1996). More than 600,000 family farming enterprises (FFEs) are involved in production. In addition to its role as a marker of land ownership in rural areas (Leplaideur 1995), the cocoa tree is increasingly a place to invest savings to secure retirement income for urban populations.

The cyclical crisis of cocoa prices in a context of state disengagement from the 1990s has led to an increase in production at the expense of quality. This has resulted in a recession in production and loss of market share, leading Cameroon to a glass ceiling of approximately 300,000 tons, today. However, the decade 2000 saw a recovery of world prices that revived production and the commitment of the state in the sector. Cameroon is the world's fifth largest cocoa producer, with production varying between 250,000 and 290,000 tons each year (Lescuyer 2019). Production, which covers 672,745 ha (FAOSTAT[1]), is concentrated in the central, southern and southwestern regions. Nearly 90% of production is carried out by small producers of 1.5–3 ha. The cultivation system is mainly based on agroforestry systems. These systems are not very intensive, with yields of 350 kg/ha. Production is mainly exported as beans. Given the current trends in the growth of world cocoa demand, Cameroon set a target of 600,000 tons in 2020, which was not reached. The new National Development Strategy Document 2020–2030 (*Stratégie nationale de développement 2020–2030* (SND-30)) sets this sector as a priority and a target of

Chapter written by Syndhia MATHÉ, Eric Joël FOFIRI NZOSSIÉ and Ludovic TEMPLE.

1 See: http://www.fao.org/faostat/fr/#data/QC.

doubling yields to 730 kg by 2030[2] (République du Cameroun 2021). It mobilizes a set of investments and technical proposals based on the desire to modernize cocoa farming practices and operations by accompanying them toward innovation mainly through the renewal of plant material, the rejuvenation of cocoa orchards and the implementation of good agricultural practices (GAPs).

This context raises the question of how to analyze and support research and innovation policies for a sustainable increase in production, that is, by reducing deforestation (Carimentrand 2020), limiting the use of pesticides and increasing the income of small producers. Carimentrand (2020) recommends the introduction of environmental sustainability issues into public action for the development of the cocoa sector. We look at the implications of this recommendation for the organization of actors who accompany innovation in the cocoa sector in Cameroon. To do this, this contribution proposes mobilizing an analytical framework in terms of the sectoral innovation system. The first part characterizes how this analytical framework allows us to identify the structure of the environment in which different cocoa-related innovations are born and evolve. It also presents the information collection and processing system. The second part presents the results in terms of identifying the dynamics of the organization of this system. The third part discusses the results in terms of research and innovation policies.

5.2. Methodological framework for analyzing information in the cocoa sector

5.2.1. *The sectoral innovation system analysis framework*

The notion of the sectoral innovation system (SIS) and production emerged in the 1990s (Breschi and Malerba 1997; Geels 2004). In the agricultural and food sector, this analytical framework was used to analyze the determinants of technological change in industry or biotechnology (Hall 2005). Its use is more recent in agriculture and agri-food, where it accounts for different possible communities of use (Touzard et al. 2014), and where the specificities of innovation processes related to the very nature of agricultural production or technological processes are questioned according to the sectors and macro-institutional contexts between OECD countries and developing countries (Casadella et al. 2017). The delimitation of an SIS in relation to these works leads to the identification of three components that interact with each other and are specified around a commodity chain or a group of products: knowledge and technologies, a system of actors operating in a network and institutions that regulate the relationships between actors (Malerba 2005).

---

2 It is articulated around eight pillars: research, production, marketing, promotion and consumption, quality, processing, financing and accompanying measures.

With regard to this reference framework, we delineate the contours of an SIS as a network of actors from the productive and institutional spheres[3] who coordinate to generate, adopt and use new technologies, new inputs (varieties, bio-inputs) or new products processed from cocoa.

Knowledge and technology are composed of basic knowledge, inputs and technologies specific to a sector. The analysis of these elements requires the specification of the sector boundaries. Actors and networks are composed of heterogeneous agents that can be firms or other organizations (research institutes, universities, financing institutions, cooperatives, trade unions, etc.). The interrelationships through formal or informal networks between these agents can be market or non-market. These agents and the networks which they are part of determining elements in identifying the boundaries of the system. Institutions include the social norms, routines, habits, practices, rules, laws and standards that influence the behavior and interactions of agents. They can therefore be formal or informal and can be deployed at different spatial scales. The dynamics of SISs are linked to two main mechanisms: a process of creation (products, technologies, organizations, institutions, behaviors, strategies) and selection by the market or other forms of non-market coordination (contracts, reciprocity, cooperation) that reduces heterogeneity within the system.

5.2.2. *Information collection system and databases*

The information collection system aims to identify the mode and logic of the organization of institutions and organizations in the cocoa sector. It combines several sources (open interviews, data mobilization database on the distribution of improved plant material, bibliographic database on research themes in Cameroon). The open interviews were conducted with organizations involved in the SIS, mainly *Office National du Cacao & Café* (ONCC), *Société de Développement du Cacao* (SODECAO), *Conseil Interprofessionnel du Cacao & Café* (CICC), *Groupement des Exportateurs du Cacao & Café* (GEX), *Fonds de Développement des filières Cacao & Café* (FODECC), the Ministries of Agriculture, Trade and Research, as well as the Industrial Cocoa Companies (SIC-CACAO) and the Chocolate and Confectionery Companies of Cameroon (CHOCOCAM).

The interviews targeted the key people within these structures who were directly involved in structuring the innovation system. The purpose of the interviews was to reveal the action strategies of the actors and the connections between these strategies polarized by the production, exchange and processing of cocoa. Different databases

3 Productive sphere: users, producers and suppliers of inputs.
Institutional sphere: universities, financial institutions, government agencies and NGOs.

were mobilized from different actors interviewed on the distribution of improved seedlings. Based on the temporal regularity of available data, we used the databases of SODECAO, MINADER and IITA[4] with comparable variables (years and intervention locations). The data considered cover the period 2009–2013. They were used to construct a spatial representation of the distribution of improved plants using a combination of Adobe Illustrator and Philcarto software. These data are considered a proxy for the levels of use of improved varieties by farmers. We also used a Web of Science bibliometric database (1991–2013) built in a study on the economics of research in Cameroon (Bisson et al. 2014). This database contains more than 7,000 scientific publications published by researchers and organizations based in Cameroon, in which an extraction of cocoa-specific publications yielded 411 references. These references reveal a diversification of scientific knowledge on cocoa over the last few decades. Finally, the authors mobilize different experiences of involvement in research projects and expertise to accompany research and innovation policies in Cameroon (Temple and Fadani 1997).

5.3. Dynamics of the innovation system in the cocoa sector in Cameroon

Agricultural policy orientations in Cameroon aim to move agriculture from an "artisanal" stage to a more modern or "second-generation" stage. In the case of cocoa production, a first recovery plan was developed in 2002, followed by a second plan from 2014 to 2020 (CTSCCC 2014). The new National Development Strategy Document 2020–2030 (NDS30) projects a production of 640,000 tons of cocoa by 2030. The two recovery plans included various actions focused on increasing yields and promoting the use of improved varieties of plants resulting from research, which have been included in the SND30. Innovation in the cocoa sector thus remains historically polarized by two complementary components.

On the one hand, is the improvement of early-growing, high-yielding plant material[5], such as the first hybrid varieties introduced with a Forastero and Trinatario dominant (Mfeck et al. 2017). Approximately 20 varieties resulting from selection carried out on research stations in the 1960s by the *Institut de Recherche Agricole pour le Développement* (Institute of Agricultural Research for Development (IRAD)) were distributed from SODECAO's seed fields. This varietal

4 MINADER: *Ministère de l'agriculture et du développement rural* (Ministry of Agriculture and Rural Development). IITA: International Institute of Tropical Agriculture.

5 Approximately 40–50% of the orchard was planted before 1950 and the average yield is 400 kg/ha.

creation dynamic faded away at the beginning of the 1990s with the state's disengagement from production activities.

Moreover, in view of the current recovery, the potential annual demand for improved seedlings is estimated by the Cocoa Development Corporation (SODECAO) at 100 million, whereas current cumulative production is only seven million[6].

On the other hand, the second component is the transfer of technical knowledge to improve cocoa farm management to fight against the main diseases and pests (brown rot and capsids) and optimize cocoa tree nutrition. Recently, and transversally, strategies for improving overall quality are being implemented to respond to the changing constraints of access to international markets (Nlend Nkott et al. 2020) concerning pesticide residues (Galani et al. 2020) or the consequences on deforestation (Carimentrand 2020).

5.3.1. *The evolution of scientific knowledge production on cocoa*

Before the end of the 2000s, scientific publications were dominated by the theme of genetic resistance of cocoa against diseases and pests (after 2008). From 2007 onward, publications diversified on agroforestry and more broadly on cropping systems. This diversification is strongly linked to the launch of the partnership facility (PF) on agroforestry[7], which has influenced the orientation of research. Despite this diversification of knowledge, two remarks are worth highlighting. The first relates to the weak presence of human and social sciences (HSS) in scientific publications, justified in particular by the absence of referencing of the knowledge produced in the Web of Science database[8].

The second reveals a weak connection between the content of the production revival projects supported by donors and public authorities, and the new production of scientific knowledge on agroforestry and sustainability in cocoa systems. In fact, the content of the production revival plans is focused on increasing production through yields, based on a "science push" model polarized by the use of inputs,

6 This accumulation includes the production of plants by all public and private stakeholders who ensure the multiplication of plant material and its distribution/marketing in all production areas.

7 The *DP Agroforestrie Cameroun* is a research platform created in 2005 that brings together CIRAD and several Cameroonian and international research and higher education institutes.

8 It should be noted that there is gray literature on the economic and social aspects, produced by researchers in the humanities and social sciences and publications in journals not referenced by the Web of Science database.

mainly new varieties. They make little or no reference to support that would enable the activation of other potential resources of socioecosystems, such as the microbiota, the capacity for endogenous innovations, or above all the complementary relationships between crops within forest systems, which would lead to the integration of spatio-ecological multifunctionality: food (Temple and Fadani 1997), ecological, forest (Seghieri and Harmand 2019) and the health of cocoa production as a resource for competitiveness.

5.3.2. *A highly administered institutional governance of innovation*

Since 1945, under the colonial administration, support for planters has been entrusted to technical organizations (Champaud 1966). This support system was based on three pillars: (1) the improvement of cultivation methods, (2) the development of plant material and (3) the control of diseases (brown pod rot, meridia and capsids). The creation of the Nkoemvone research station[9] in 1949 set the framework for future scientific research. This structure will later become the specialized station of IRAD, the crucible of the improvement and multiplication of cocoa varieties.

The objective of increasing production was supported by several public structures over time. The "*Secteur Expérimentale de Modernization des cacaoyères*" (SEMAC), created in 1953–1954, was made up of several "peasant posts" with the mission of providing supervision and extension. The "Services de l'agriculture", ancestors of the Ministry of Agriculture, reinforced the supervision of farmers through the "Postes agricoles". The *Société de Développement du Cacao* (SODECAO), created in 1974, inherited the organizational system and missions of SEMAC. A Department of Phytosanitary Control (LPAC) completed this public system.

The mapping of the network of actors that structure the administered sector thus reveals a concentration of institutional action on the production segment (Figure 5.1). This choice can be justified by the objective of increasing production in a context of strong competition between producing countries. In addition, Cameroon's public action in the area of production was financed by two sources: first, national mechanisms for regulating the sector, which led to the creation of the "Stabilization Fund" in 1955 and the "National Commodity Marketing Office" (*Office National de Commercialisation des Produits de Base* (ONCPB)) in 1976; and second, international mechanisms for stabilizing exports (STABEX Fund, 1975–1999). Domestic marketing and export, although theoretically in the hands of the private sector, was controlled by the public sector (Laporte 1992; Folefack 2003).

9 Locality located in the Mvilacocoa basin (see Figure 5.3).

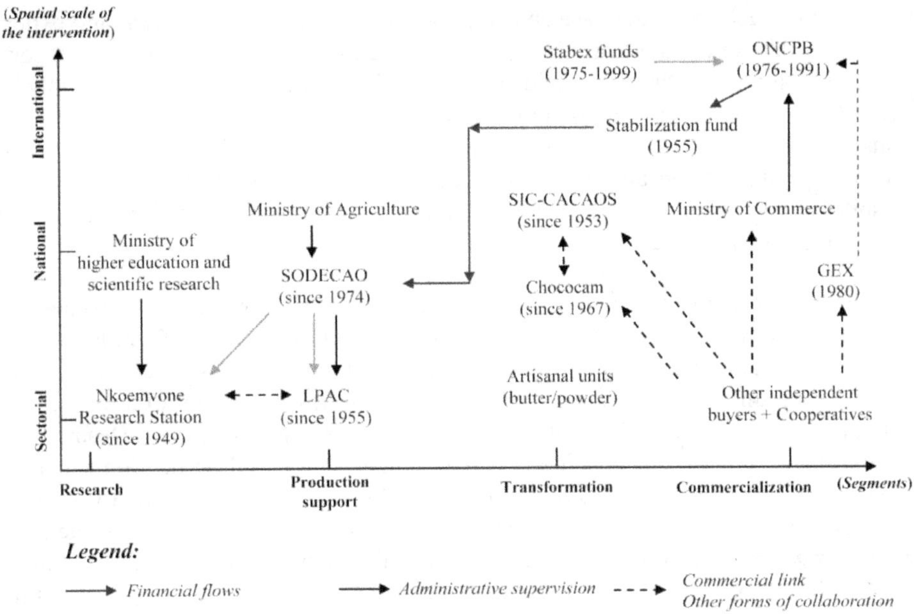

Figure 5.1. *Stakeholder system mapping for the cocoa sector under administration. For a color version of this figure, see www.iste.co.uk/casadella/agri-innovations.zip*

This mapping reflects the public will to specialize and ensure complementarity between the different organizations that make up the production support subsystem in the value chain. This complementarity also constituted an important vector for the transfer of scientific and technical knowledge to the planters (cultivation, plantation maintenance, disease control, harvesting, drying and storage) through SODECAO, the secular arm of the state. This subsystem also integrated a multiscalar territorial dynamic of productive and commercial structuring (sectoral, national and international), corresponding, respectively, to production zones, purchasing points for farmers and international markets. This control of the cocoa space at different scales has made it possible to efficiently constitute a reliable statistical base on production (number of farmers, cultivated areas, tonnages) and domestic and international marketing.

5.3.3. *"Liberalization": a vector of sectoral transformation of innovation in cocoa*

The reform of the cocoa sector at the end of the 1980s diversified the system of actors (Figure 5.2). It was marked by the disengagement of the state from the production structures at the base and from the supervision of the internal and export marketing network. Four specific sectoral segments (research, production support, processing and marketing) emerged from this reform, which broke the monopoly of SODECAO. Thus, the elimination of public subsidies (1987–2001) led to the transfer of some of its missions to the Ministry of Agriculture (supervision of growers, multiplication and distribution of plant material, protection of orchards). Despite the recent rehabilitation of SODECAO, most of its missions are still carried out by ministerial projects.

The appropriation of these missions raises questions about the effectiveness of statistical monitoring of production, expertise and adequate infrastructure for orchard renewal and protection against diseases. In the framework of the anti-capsid control program previously carried out by SODECAO, the creation of a planter file structured the construction of a national statistical database. Despite its rehabilitation, the SODECAO file covers less than 80% of all production basins at the national level. Available statistics on national production come from a variety of sources (Ministry of Agriculture, SODECAO, ONCC, GEX). They sometimes contain significant discrepancies, even though they can be cross-checked through exports by integrating the proportion processed at the national level. Processing (approximately 20% of production) is controlled by the *Sociétés Industrielles des Cacaos* (SIC-CACAOS), a subsidiary of the Barry Callebaut Group, and the *Chocolateries et Confiseries du Cameroun* (CHOCOCAM). As part of the revival of the sector, the construction of a dozen processing units will increase the proportion of processed cocoa to nearly 30% of current production.

The reform of the sector has redefined the responsibilities of the various stakeholders in the sector. It disengages the state bodies from marketing. The state, through the "National Cocoa and Coffee Board" (ONCC), which replaced the ONCPB in 1991, assumed the functions of regulator. The key players in the sector (producers, buyers, industrial processors and export professionals) have been grouped together since 1991 within the Interprofessional Council of Cocoa and Coffee (CICC). Cocoa exports are managed exclusively by the private sector, represented by the Group of Cocoa-coffee Exporters (*Groupement des Exportateurs de Cacao-café*, GEX), which was created in 1980 and has approximately 60 members. A system for financing institutional actors and public projects in the cocoa and coffee sectors was also established in 2006 with the creation of the Cocoa and Coffee Development Fund (FODECC).

Liberalization has thus led to an increase in the number of stakeholders and a more complex relationship between state regulation and private sector actors. Cocoa is at the heart of antagonism between public organizations, on the one hand, and between public and private structures, on the other hand, questioning the consequences on the coherence of research, innovation and support policies for the sector.

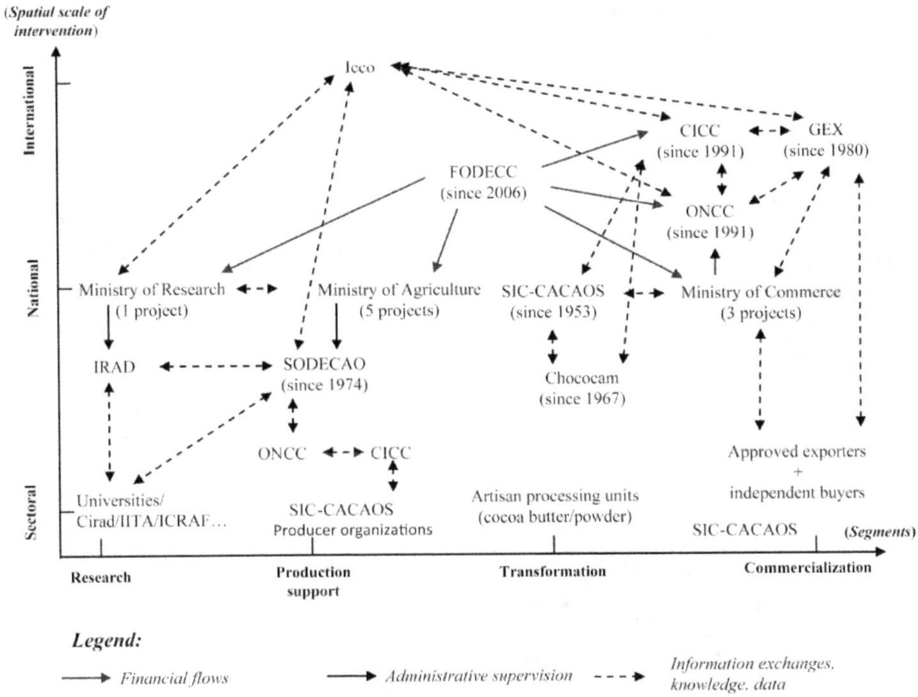

Figure 5.2. *Mapping of institutions structuring the sectoral innovation system*

A key function at the heart of this antagonism concerns the production and distribution of improved varieties. This function is currently undermined by individualities involving state bodies (Ministry of Agriculture, Minader, SODECAO, Fodecc), private organizations and companies (CICC, SIC-CACAOS), research institutions (IITA, IRAD, CIRAD) and political elites. Analysis of the mobilizable data from the three main stakeholders (Sodecao, Minader and IITA) shows interannual variations that raise questions about the consolidation of SODECAO's leading role as the secular arm of the state. However, from a spatial point of view, the distribution of improved seedlings reflects a strong demand from producers for the rejuvenation of their orchards in traditional cocoa basins and for the development of pioneer fronts, particularly in Mbam (Figure 5.3).

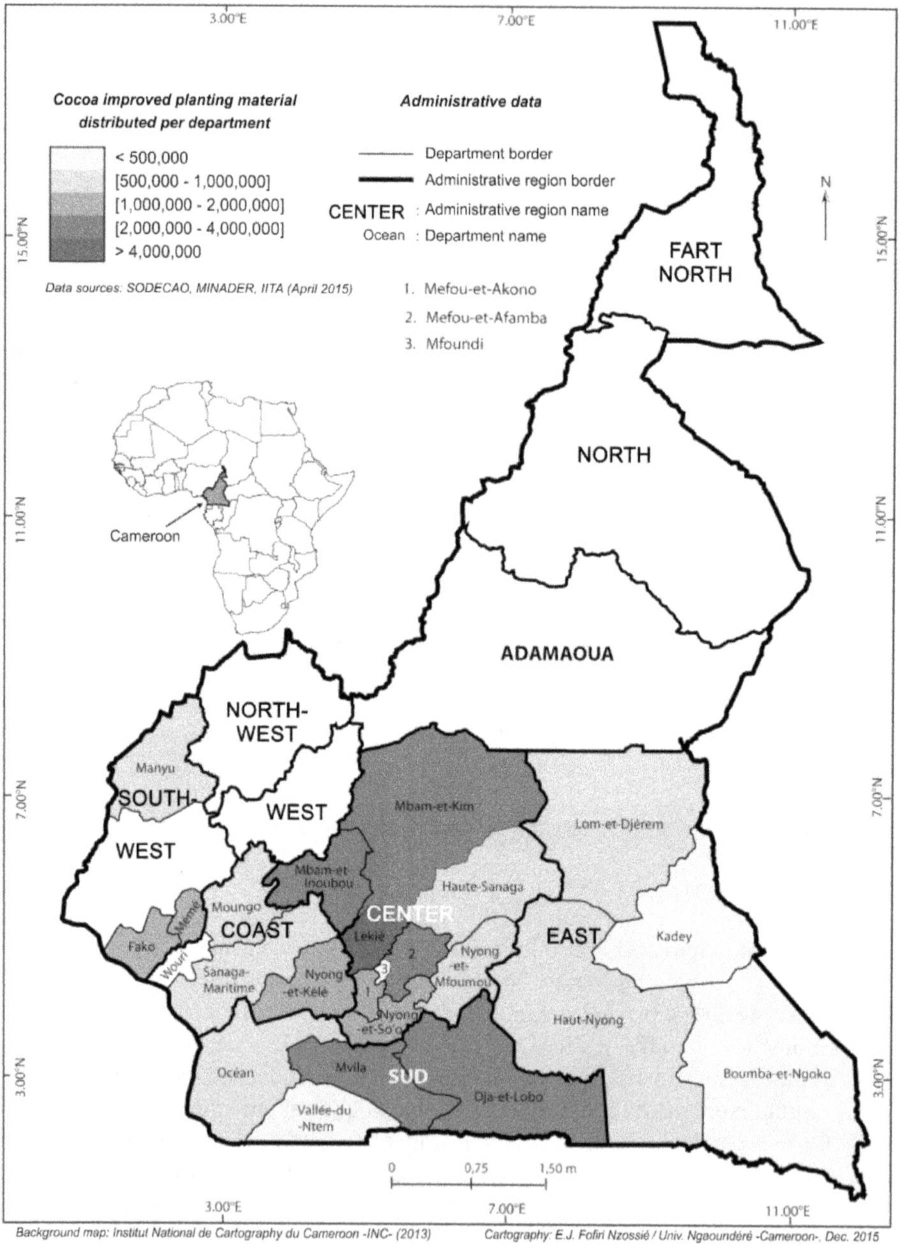

Figure 5.3. *Distribution of cocoa seedlings from improved planting material in the production basins in Cameroon (2009–2013)*

Another set of key functions related to sustainability is that of certification (Nlend Nkott et al. 2020), which began to be implemented in Cameroon in the 2010s through different standards. The objective of the evolution of the specifications is mainly to reduce deforestation. It translates into the recommendation of good agricultural practices based sometimes on a higher use of pesticides. The actors involved in certification are mainly NGOs, producer organizations and cooperatives and private companies (exporters) (Nlend Nkott et al. 2020).

5.4. What are the consequences for the sector's research and innovation policies?

5.4.1. *Variable governance of the sectoral innovation system questioned by agroforestry*

Analysis of the cocoa sector system in Cameroon shows two main visions for the development of this sector. The first is driven by the predominance of "actor research" in the innovation system. In the past, cocoa research has been very much oriented toward biophysical aspects, with little consideration given to basic knowledge on economic, institutional and organizational aspects or focused on microeconomic analyses (Leplaideur 1995). This technical bias allowed little or no reference to the meso- and macroeconomic constraints on innovation of a social, economic, organizational and institutional nature.

On the other hand, in recent years, the recent orientation of agronomic research on agroforestry and the intensification of "cropping systems associated with food crops" has broadened the knowledge base and the processes of innovation that they are allowed to document. Thus, work on the conditions of land development associating trees, food crops or animals simultaneously or sequentially implies resituating cocoa production in the overall dynamics of the agrarian system. The idea is to improve the ecological and economic interactions between the associated forms of development. This approach favors the search for conditions to maintain the productivity of the soil on cultivated land rather than the exploitation of new land, while at the same time diversifying the technical model for the different ways of increasing production. It makes it possible to reconcile the health of plants, ecosystems and farmers with less use of synthetic inputs.

On the other hand, there is the vision carried by various institutional actors and downstream companies in the sector, with a desire to increase production through productivity based on research results in an innovation model based on the "science push". This vision can be explained in three dimensions. The first is that of a certain nostalgia for the prosperous period of state supervision in the face of current institutional inertia. The second is the upward trend in cocoa prices on the international market. The third is the incentives given by international donors. In this context of recovery, production is largely driven by international prices, and the supply of seedlings and seeds from research is mainly achieved through the extension of pioneer fronts for forest or sometimes savannah development (as in the Mbam plain).

The sectoral policy of steering innovation through the dissemination of cocoa planting material through its specialized bodies (SODECAO, ONCC, FODECC), but also through parapublic (CICC) and private (GEX, SIC-CACAOS, CHOCOCAM) organizations and enterprises is being questioned by new research that diversifies the conditions for designing innovation. Indeed, the suitability of these varieties to the objectives of optimizing complex ecosystems within the framework of associated systems, adapted to localized situations and anticipating the needs for adaptation to climate change and the depletion of the soil fertility rent on a perennial plant, questions the relevance of current orientations.

The revival of the sector is a development issue for Cameroonian agriculture. This issue has strong economic and financial interests, particularly in the production and marketing segments. Cocoa is one of the main sources of value for companies in the global agri-food sector and the agro-chemical sector (exporters, importers, chocolate manufacturers, phytosanitary products, etc.). The institutional weight of these actors in the current technological orientations is major with regard to the disconnections observed between the structure of current development actions and the research offer.

In the production segment, which is the focus of the recovery plan projects, there is little coordination and sometimes competition between technical organizations for the extension of research results, sometimes leading to a duplication of roles and a lack of coordination. The existence of this competition is explained by the history of the construction of the institutional system of production supervision, which, in the process of liberalization, has seen new actors superimposed on the same functions, particularly that of distributing cocoa improved seedlings, but increasingly supervising producers. Coordination could initially focus on reuniting the statistical information system for monitoring orchard renewal. In the current state, the diversity of statistical sources creates opacity on the correct calibration of seedling supply needs, but also a risk for understanding the solutions in the case of a health crisis affecting the cocoa orchard. Finally, the integration of monitoring the environmental and social impacts of new varieties seems equally important for the proper orientation of research breeding programs. This should be done with regard to the objectives of systemic sustainability and not only with regard to the yields obtained at the experimental station.

5.4.2. *A challenge of innovation governance through the hybridization of scientific and local knowledge*

Knowledge, not only of a technical nature, may or may not be specific to the sector. It is conveyed through learning and experimentation processes within the sector to the extent that it is useful, usable and used. It can come from research, as well as from non-academic actors (private companies, civil society, rural communities). If knowledge is presented as a key element of SISs, it must be specified that it is not the stock of knowledge itself that is important, but rather its circulation between the actors of the system and the collective learning processes that test its appropriation and use in the implementation of changes (Brette 2014). This circulation requires that this knowledge be translated for different communities of actors, and thus transformed and relayed by intermediary actors sometimes called "brokers" (Klerks et al. 2009, 2012). Initially, this translation was carried out by technical organizations under the supervision of the state, mandated to promote research results. Since the liberalization policies, these intermediation structures have diversified via projects supported by international cooperation in the private sector, mainly NGOs. The latter sometimes find themselves in competition, not only with each other, but also with public organizations. However, despite this renewal of intermediation structures, we note a loss of momentum in the transfer logic, which is potentially accounted for by the low adoption by producers of technologies proposed by research (Duchamp et al. 2014; Fofiri et al. 2015).

The proposed technologies either do not meet or no longer meet the needs of producers and the market, or their use proves too restrictive or risky given the conditions of access to farmers' resources, their vulnerability and its increase (climatic and health crises). This loss of momentum explains why approaches in terms of innovation systems attempt to promote more collaborative innovation models based more on the hybridization of farmers' knowledge bases and scientific experiments than on simple technological transfers. The latter are all the more important in developing countries where information systems and formal training are not easily accessible to farmers in precarious situations, and where access to codified knowledge and its products is only available to elites in particular situations. Thus, improving innovation systems by taking into account the knowledge gained from experience and collective learning, which is often not codified in technical data sheets or research articles, is an important issue for strengthening ISS.

5.4.3. *A potential reconfiguration of innovation models in the cocoa sector*

The linear innovation model of technology transfer of high-yielding varieties from research to producers still underpins, in part, the innovation policies structured by the current production revival projects supported by public authorities and donors. These projects continue to focus on the dissemination of new inputs (varieties, pesticides). The new, more systemic models of innovation, which refer to a more complex approach, require different policy orientations (Edquist and Zabala 2012). They imply a better articulation between actors in the design of innovation policies (Temple and Casadella 2020) that are structured by the development projects supported, the knowledge proposed by research, and the knowledge bases and experiences of farmers linked to collective actions based on the resolution of localized constraints.

This broadening requires us to stop focusing on a single perspective of responding to a need for rapid growth in production that is driven at the global level by various bodies, and even by the demand of globalized firms. It requires the integration of questions relating to the environmental and social sustainability of production (Carimentrand 2020), and structures three proposed recommendations.

The first is to strengthen the coordination of actors involved in the cocoa sector by clarifying, or even redefining their roles to avoid overlapping the same functions and the same targets. In the current context of recovery, this coordination can be organized within an interinstitutional cell such as the Technical Cell for Monitoring and Coordination of the cocoa and coffee sectors. It would involve, for example, reframing the role of research institutions in the production of knowledge and outputs (improved planting material, Good Agricultural Practices), and agro-industrial companies in the formulation of quality improvement needs and the support of production dynamics (financing of inputs).

The second proposal is to support the expansion of the capacity of research to take into account the specificity of the constraints (social and economic organizations, land issues) that structure sustainability (a vector of competitiveness in the long term). This includes questions on the adaptation of technical choices to the needs of optimizing ecosystem resources, reducing dependence on the chemical industry, adapting to climate change and improving the conditions for distributing value for the most vulnerable categories of actors, including planters. Although there has been a significant increase in the production of scientific knowledge on plant interactions within agroforestry systems (Jagoret et al. 2011), on the epidemiology of diseases in relation to different ecosystems (Hall 2005), and on the impact of climate change (Magne et al. 2014), the use of this scientific knowledge by the organizations in charge of the technical revival of cocoa production remains to be confirmed. It is therefore necessary to consider the establishment of a national structure for dialog on intermediation between research results and production actors to clearly evaluate and structure the necessary infrastructure to meet the needs of production actors.

The third proposal is to increase investment in supporting the professionalization of producers, structuring private actors at the heart of intermediation between exporters and growers, and strengthening their managerial capacities to guarantee the level of income of producers in their bean marketing relationships.

5.5. Conclusion

Through the mobilization of the SIS analytical framework, we qualified the trajectories of innovations in the cocoa sector in Cameroon by identifying the institutional, historical and political conditions of knowledge and technology production. The exploratory use of this framework, however, remained partial with respect to the information and knowledge systems that could be mobilized. The results nevertheless document different options for reconfiguring public support and research policies for a cocoa sector that integrates other parameters related to the sustainability of productive socioecosystems into the search for competitiveness based on yield (Lescuyer et al. 2019). They currently highlight a real need for coordination and reorganization of the institutional framework for production. Maintaining the classic objective of focusing research and innovation on the quantitative increase in production (increase in yields and planted areas) is not sufficient for sustainable competitiveness. World cocoa market cycles are subject to reversals over time. The integration of variables related to the optimization of ecosystem resources, the reduction of pesticides, compatibility with food crops and adaptation to the effects of climate change in cocoa production development projects are constructive orientations for a sustainable cocoa industry in quality niches, which increase the capture of value in the global sector. The public institutionalization of cocoa certification systems is a lever for the governance of technical innovations that are mobilized in the cocoa sector. It promotes good agronomic practices in the elaboration of the specifications that underlie these certifications, which will make it possible to truly reduce the use of pesticides. It imposes new forms of collaboration between the different actors in the innovation system and challenges linear conceptions of innovation based on the simple dissemination of technical innovations developed in research laboratories.

5.6. References

Alary, V. (1996). Incertitude et prise de risque en période d'ajustement : le comportement des producteurs de cacao du Cameroun avant et après 1994. PhD Thesis, Université de Paris-Panthéon Sorbonne.

Bisson, P., Alami, S., Temple, L. (2014). Etude sur l'économie de la recherche au Cameroun. Programme d'Appui à la Recherche. Ministère de la recherche scientifique et de l'innovation. Document, Cirad, Yaoundé.

Breschi, S. and Malerba, F. (1997). Sectoral innovation systems: Technological regimes, Schumpeterian dynamics and spatial boundaries. In *Systems of Innovation: Technologies, Institutions and Organizations*, Edquist, C. (ed.). Frances Pinter, London.

Brette, O. (2014). La valorisation économique de la recherche scientifique. In *Principes d'économie de l'innovation*, Boutillier, S., Forest, J., Gallaud, D., Laperche, B., Tanguy, C., Temri, L. (eds). De Boeck, Brussels.

Carimentrand, A. (2020). Cacao. Etat des lieux sur la déforestation et les standards de durabilité. Report, CST Forêts de l'AFD and Cirad, Montpellier.

Casadella, V., Tahi, S., Temple, L. (2017). Créativité et politiques d'innovation. *Journal of Innovation Economics & Management*, 53(2), 5–12.

Champaud, J. (1966). L'économie cacaoyère du Cameroun. *Cah. Orstom*, 105–124.

CTSCCC (2014). Plan de relance et de développement des filières cacao et café du Cameroun Horizon 2020. Cellule technique de suivi et de coordination des filières cacao et café (CTSCCC), Services du Premier Ministre, Secretariat Général, Cameroon.

Duchamp, M., Temple, L., Ndoumbe Nkeng, M. (2014). Cadrages méthodologiques de l'analyse bibliométrique. Etude sur l'économie de la recherche au Cameroun. Report, Minresi-Cirad, Yaoundé.

Edquist, C. and Zabala-Iturriagagoitia, J.M. (2012). Public procurement for innovation as mission-oriented innovation policy. *Research Policy*, 41(10), 1757–1769.

Fofiri Nzossie, E.J., Temple, L., Ndjouenkeu, R. (2015). La contribution de la recherche universitaire à la formation d'un système sectoriel d'innovation agro-alimentaire au Cameroun. *Innovations*, 47(2), 55–77.

Folefack, D.P. (2003). Analyse socio-économique de la commercialisation du cacao en zone forestière du sud Cameroun. Diplôme d'études spécialisées en économie et sociologie rurales, Faculté Universitaire des Sciences Agronomiques de Gembloux.

Galani, Y.J.H., Houbraken, M., Wumbei, A., Gjeugap, J.F., Wumbei, A., Fotio, D., Gong, Y.Y., Spanoghe, P. (2020). Monitoring and dietary risk assessment of 81 pesticide residues in 11 local agricultural products from the 3 largest cities of Cameroon. *Food Control*, 118, 107416.

Geels, F.W. (2004). From sectorial systems of innovation to socio-technical systems: Insights about dynamics and change from socilogy and institutional theory. *Research Policy*, 33(6–7), 897–920.

Hall, A. (2005). Capacity development for agricultural biotechnology in developing countries: An innovation systems view of what it is and how to develop it. *Journal of International Development*, 17(5), 611–630.

Jagoret, P., Michel-Dounias, I., Malezieux, E. (2011). Long-term dynamics of cocoa agroforests: A case study in central Cameroon. *Agroforestry Systems*, 81(3), 267–278.

Klerks, L., Hall, A., Leeuwis, C. (2009). Strengthening agricultural innovation capacity: Are innovation brokers the answer? *International Journal of Agricultural Resources, Governance and Ecology*, 8(5), 409–438.

Klerks, L., Schut, M., Leeuwis, C., Kilelu, C. (2012). Advances in knowledge brokering in the agricultural sector: Towards innovation system facilitation. *IDS Bulletin*, 43(5), 53–60.

Laporte, B. (1992). Les réformes des systèmes de commercialisation et de stabilisation des café et cacao au Cameroun et en Côte-d'Ivoire. Article, CERDI, Clermont-Ferrand.

Leplaideur, A. (1995). Stratégies paysannes autour des systèmes cacaoyers et vivriers en zone centre-sud forestière camerounaise. In *Etats, développement, paysans : actes du colloque*. CIRAD-MESRU, Montpellier.

Lescuyer, G., Boutinot, L., Goglio, P., Bassanaga, S. (2019). Analyse de la chaîne de valeur du cacao au Cameroun. CIRAD, Wagenigen University & Research.

Magne, A.N., Nonga, N.E., Yemefack, M., Robiglio, V. (2014). Profitability and implications of cocoa intensification on carbon emissions in Southern Cameroun. *Agroforestry Systems*, 88(6), 1133–1142.

Malerba, F. (2005). Sectoral systems of innovation: A framework for linking innovation to the knowledge base, structure and dynamics of sectors. *Economics of Innovation and New Technology*, 14(1–2), 63–82.

Mfeck Eyenga, L.B., Kamdem, C., Temple, L., Mathé, S. (2017). Rendements et mécanismes d'adoption du matériel végétal amélioré : le cacao au centre-Cameroun. *Tropicultura*, 35(2), 110–120.

Minkoua Nzié, J.R. and Temple, L. (2018). L'offre de recherche converge-t-elle vers les besoins du secteur agroalimentaire au Cameroun ? Une analyse par la bibliographie. *Cah. Agric.*, 27, 25008.

Nlend Nkott, A.L., Mathé, S., Temple, L. (2020). Analyse multi-niveau des freins à l'adoption de la certification du cacao au Cameroun. *Economie rurale*, 370, 81–99.

République du Cameroun (2021). Stratégie de développement du secteur rural/plan national d'investissement agricole SDSR/PNIA (2020–2030). Report, République du Cameroun.

Seghieri, J. and Harmand, J.M. (2019). *Agroforesterie et services écosystémiques en zone tropicale*. Quae, Versailles.

Temple, L. and Cassadella, V. (2020). Politiques et modèles d'innovation en Afrique : enjeux systémiques et structurels par le renforcement des capacités. *Technologie et Innovation*, 5(3) [Online]. Available at: https://www.openscience.fr/Politiques-et-Modeles-d-Innovation-en-Afrique-Enjeux-systemiques-et-structurel.

Temple, L. and Fadani, A. (1997). Cultures d'exportation et cultures vivrières au Cameroun : l'éclairage d'une controverse par une analyse micro-économique. *Economie rurale*, 239, 40–48.

Temple, L., Machikou Ndzesop, N., Fongang Fouepe, G.F., Ndoumbè Nkeng, M., Mathé, S. (1997). Système national de recherche et d'innovation en Afrique : éclairage du Cameroun. *Innovations*, 2(53), 41–67.

Touzard, J.M., Temple, L., Faure, G., Triomphe, B. (2014). Systèmes d'innovation et communautés de connaissances dans le secteur agricole et agroalimentaire. *Innovations*, 43(1), 13–38.

6

Valorization of the Date Industry in Tunisia by Combining "Modern" and "Traditional" Knowledge and Techniques: Difficulties, Successes and Prospects

6.1. Introduction

Phoeniculture, another name given to date palm cultivation, is a fundamental agriculture in some Middle Eastern and North African producing countries. It has been cultivated in arid or semiarid lands for thousands of years. Today, the world production of dates has quadrupled in 40 years and represented, in 2020, more than 9 million tons per year (Foprodatte 2022). The main producers are Egypt (19%), Saudi Arabia (14%) and Iran (14%). Tunisia, despite its 4% of world production in 2020, ranks as the leader in date exports in value (export volume measured by total turnover generated) with 19% of world exports. Its global leadership remains fragile, however, due to very strong international competition. In this regard, Tunisia lost its first place position in 2022 to Saudi Arabia. In addition, the consumption of dates, particularly appreciated during the festive periods of Ramadan and Christmas, is widespread on all five continents of the world to varying degrees. Twelve varieties of dates exist, but "Deglet Nour" (12% of world production) and "Medjoul" (1% of world production) are particularly appreciated by consumers. The variety "Deglet Nour" is largely grown in the southern regions of Tunisia.

Countries whose economies are based on a symbolic agriculture of a fruit or vegetable, such as avocado growing in Kenya (Mwanbi et al. 2016) or horticulture in Vietnam (Mergenthaler 2009), require a global strategic reflection to ensure the

Chapter written by Yasmine BOUGHZALA and Néfissa BEN MAHMOUD.

viability and sustainability of its value chain. In Tunisia, agriculture represents nearly 11% of GDP through strategic sectors such as olive oil, dates and citrus fruits. Today, the agricultural sector is undergoing considerable technological transformation through Agriculture 4.0 (Arora et al. 2022; Sharma 2022) and the various innovations it is experiencing (Cagliano et al. 2016). Agricultural sectors also face major challenges such as environmental threats related to climate change or water stress. The economic and political stakes are also considerable. Paradoxically, they vacillate between increasingly competitive international competition for certain agricultural products and political resolutions aimed at food sovereignty. Recent international crises such as the one in Ukraine show the seriousness of its impact on the world grain market.

These various economic, social and environmental issues are of concern to both researchers and practitioners. This chapter examines the development of the date industry in Tunisia. Indeed, it will first examine what the guarantors of its world leadership are in terms of exports in value, cruelly essential to its local economy. Secondly, we will analyze the extent to which this thousand-year-old value chain is adapting to the technological and environmental transformations commonly impacting the world economies of our time. To do so, we will present a first section dedicated to a literature review on agricultural value chains in emerging countries, then the methodology adopted and finally the results observed and presented in the form of challenges to be taken up by organizations involved in this sector as well as local public authorities.

6.2. Literature review

6.2.1. *Sector and value chain: complementary but distinct concepts*

Firstly, we feel it is important to revisit the concepts of sector and value chain (henceforth VC) to bring greater clarity to our analysis. These two notions are often ambiguous and analogous because they refer to two similar concepts but do not usually have the same use or the same definitions. Traditionally, the concept of sector refers to a homogeneous set of economic activities linked vertically and horizontally by commercial exchanges (Shaffer 1973). It can be divided into four main groups: production, processing, marketing and consumption (Carimentrand 2012; Ranomenjanahary 2018). For the value chain, we retain the definition of Gerefi and Fernandez-Stark (2016) according to which:

> The value chain describes the set of activities that companies and workers perform to take a product from conception to end use and beyond. It includes activities such as research and development (R&D), design, production, marketing, distribution, and support to the end

consumer. The activities that make up a value chain can be contained within a single company or spread across different companies.

This definition emphasizes the optimization of the firm's relationships with the strategic actors who chain their activities from upstream to downstream (Gerefi et al. 2005) while stressing the importance of mobilizing relationships with the set of services (research-innovation, transportation, training) and the interconnection of markets that transform inputs into products. It allows the needs of consumers to be met and ensures better competitiveness (Lançon et al. 2017; Devaux 2018).

6.2.2. *Agricultural value chains in emerging countries*

Agricultural value chains are often at the heart of the economies of some emerging countries. In a fiercely competitive environment, the globalization of production has allowed for the rapid integration of several emerging countries into the global economy (Pananond 2022). This integration has disrupted global trade and global value chains through the steady increase in their competitive power (Trepant 2008). The term emerging market economy was coined by Van Agtmoel in 1981. They are characterized by significant and rapid economic growth with economic openness to the rest of the world. This is reflected in an increase in overall and per capita gross domestic product (GDP) and an increase in the volume of trade through significant exports (Trepant 2008; Khanna and Palepu 2010; Carrasco et al. 2012). Other authors define emerging countries by their most important characteristics. According to Cavusgil and Ghauri (1990), countries with weak industrial and social structures that make economic reforms and programs to solve the problems of unfair income distribution and unemployment can be defined as emerging economies. Therefore, they are generally the so-called BRICS countries (Brazil, Russia, India, China, South Africa) and more recently the MINT (Mexico, Indonesia, Nigeria and Turkey).

Typically, through their participation in global value chains, firms in emerging countries have the opportunity to access critical information about the type and quality of products required by consumers in the international market (Gerefi 1999; De Marchi et al. 2018; Taglioni and Winkler 2018).

6.2.2.1. *International challenges*

Global value chains differ from traditional production and trade in a few key characteristics, such as customization of production and sequential production decisions from buyer to suppliers (Taglioni and Winkler 2016; Jones et al. 2019). Recently, with the rise of health crises, social changes and environmental impact awareness, global high-value buyers demanding agricultural products have become more attentive to food quality and safety (Liu et al. 2019; Pappa et al. 2019). As sustainability holds increasing importance and figures as the new value component, agricultural GVC links face several economic, environmental and social challenges (Manfroy 2021). They must therefore adapt to new business and regulatory issues, including industrial performance and food safety, to meet market demands (Humphrey and Memedovic 2006; Devaux et al. 2018; Pappa et al. 2019; Lezoche et al. 2020).

According to Humphrey and Memedovic (2006), agricultural value chains face the following three main challenges:

a) the growing importance of international standards: meeting the food safety requirements of importing countries has become more complex as standards and compliance tests increase;

b) the requirement of global agricultural buyers: these requirements may include high volume supply, speed and reliability of delivery, customization of products through processing and packaging, and guarantees on the safety of agricultural products;

c) valuation and differentiation of agricultural products: generally, the added value of these products lies in the provision of one or more certifications or quality labels such as BIO, Demeter, Ecocert, Fair Trade, ISO 9001 (Schoneveld et al. 2019) or through closer links with traders, processors or retailers. Product distinctiveness must be established at the point of origin and maintained throughout the value chain.

6.2.2.2. *Social and environmental challenges*

At the level of emerging countries, farmers are usually the weakest link in an agricultural value chain. Indeed, they usually face several challenges when trying to develop sustainability practices (Schoneveld et al. 2019; Hidayati et al. 2021a, 2021b). They are usually trapped in a traditional system characterized by limited resources such as low productivity, inconsistent quality, limited education and limited access to market information, which prevents them from achieving increasing compliance requirements (Schoneveld 2019). In an international context, sustainable AVCs bring better wages for farmers and generate rural poverty reduction, especially since more than half of these working populations find employment in agriculture (Chotia and Rao 2017; Manfroy 2021).

This sustainability is achieved, for example, through the use of “Fair Trade”, which is the most recognized and reliable sustainability label in the world. This certification allows buyers of agricultural products to be certain that during the AVC process, there has been an implementation of decent working conditions, a ban on discrimination and child labor, in addition to a guarantee that purchase prices aim to cover the average costs of sustainable production of these products (Fairtrade.net). Several companies have strongly emphasized their models based on the direct purchase of certified agricultural products for their sustainability and contribution to better living conditions of small farmers involving better value addition (Ingram et al. 2016; Lescuyer 2020), such as Starbucks and Ahold.

In some agricultural value chains, such as the cultivation of coffee around the Luki Biosphere Reserve (LBR) or the cocoa AVC in Cameroon, workers face significant health risks. The impacts are limited to the production of coffee around that region since no chemicals are used for the maintenance of the plantations. Secondly, the processing link responsible for manual roasting provides backbreaking work, and workers are exposed to high heat and a large amount of dust in factories without having protective equipment (Manfroy 2021). In the case of cocoa cultivation, the tools used to apply fertilizers and insecticides are handled without precautions (masks, gloves, boots, etc.) and by children. This is because the cost of safety equipment is high for small farmers (Lescuyer 2020).

At the environmental level, the traditional cultivation of agricultural products has many negative impacts on the environment and biodiversity of ecosystems, including CO_2 emissions, deforestation, land, soil and forest degradation, depletion of non-renewable resources and others (Ingram et al. 2016; Gaveau et al. 2017; World Bank 2020). In this context, the consequences of the cultivation of certain agricultural products have been the subject of real awareness by the different actors of the value chain. They are increasingly moving toward more climate-friendly international standards and mandatory sustainability reporting regimes (Taglioni and Winkler 2016; Amiel et al. 2019; Manfroy 2021), such as the extension of certifications, pressure from Western markets to have sustainable agricultural products and incentive bonuses for each kilo of cocoa sold to companies committed to environmental sustainability (Lescuyer 2020). As an example, we can cite Nestlé, which developed a "cocoa plan" in 2019, in which it commits to act against deforestation by implementing measures to exclude suppliers in case of illegal deforestation, training producers and promoting access to credit (Amiel et al. 2019; Nestlé 2019).

6.2.3. *The date industry in Tunisia*

6.2.3.1. *The date, fruit of paradise*

The date palm (Phoenix dactylifera) is a majestic tree symbol of the civilizations of the Arab world that grows in hot semiarid, arid to hyperarid regions. Date palms, with their ability to adapt to desert environments, extreme temperatures and lack of water, have given mankind the opportunity to settle in the middle of hostile desert areas. They helped nomadic populations organize communities and learn about agriculture (El Hadrami and Al-Khayri 2012). It represents a particular icon of agriculture in the MENA region encompassing the Near East and North Africa (Al-Yahyai and Manickavasagan 2012; Al-Alawi 2017). The date fruit consists of a fleshy pericarp and a stone (seed) that has 7–30% of its weight. The seed varies in size from less than 0.5 to approximately 4 g, its length from approximately 12–36 mm and width from 6 to 13 mm, and it is characterized by an albumen that is white, hard and horny, protected by a cellulosic envelope (Espiard 2002; Elbana 2018). This fruit is a very beneficial food for health. Indeed, it is marketed worldwide as a fruit possessing nutritional and therapeutic value with antioxidant, antibacterial, antifungal and antiproliferative properties (Al-Yahyai and Manickavasagan 2012; Al-Alawi 2017). From the point of view of the production process, good quality dates do not emanate by chance. In fact, in addition to good climatic conditions and suitable palm grove areas, the effort exerted by the production organizations to have good quality dates is crucial from harvesting to marketing.

6.2.3.2. *Date propagation techniques*

It seems important to trace the important phases characterizing the agricultural value chain of the date, starting with its genesis (the seed), through cultivation and production and finally its marketing.

The date palm is dioecious. In other words, it has separate male and female plants. Accordingly, there are three propagation techniques for this tree: propagation by seed, propagation by branches that appear mainly near the base of the stem in the first years of the palm's life (traditional methods) and tissue culture techniques (Zaid and De Wet 1999).

In this section, we will discuss each propagation technique.

Firstly, dates are easily grown from seeds. However, dates from seedlings are often smaller and of lower quality (Elbana 2018). This technique is also called sexual propagation, and while it is useful, it is not a practical method. In fact, knowing that date palm is a dioecious species, half of the offsprings will be male and half will be female. It is not possible to determine at an early stage the sex of the offspring or the quality of the fruits or pollen before flowering, which usually takes place 7 years later. Additionally, planting seedlings rarely produces good-quality fruits with the same desirable characteristics. This is due to the heterozygous character of the date palm, which does not allow for true-to-type propagation. Finally, at the marketing level, it will be very difficult to collect the dates and market them as a single crop, hence the loss of money. However, the Saudi company Bateel uses the method of pollination through human assistance. We carefully place branches of white male flowers among the female flowers in a process that requires constant care and attention. Only the female trees produce fruit, and once pollinated, they begin to produce fruit after six years (Bateel official website).

Then, the second method of propagation is by removing and planting the branches that appear near the stem during the first years of the adult date palm's life from its fourth or fifth year once a year. In fact, palm can easily produce two branches if the soil conditions, irrigation and fertilization are good (Tengberg 2009; Time Is 2019). The technique of rejection or branching is recommended because it is easy and involves less work, time and costs; in addition, the resulting plants are faithful to the parent palm type.

Finally, the tissue culture technique is also called the in vitro technique. As the name suggests, date palms are grown in laboratories. This technique has many advantages since the varieties are well chosen based on their exemption from diseases and pests and other characteristics. Moreover, multiplication is performed on a large scale and produces uniform plants.

6.2.3.3. *Evolution of traditional production toward a sustainable production of dates*

Once the seed or branches of the date palm are anchored in the soil, farmers begin their production process. In this section, we will focus on the aspect of sustainability integrated into agricultural value chains. Recently, the agribusiness sector has faced compelling environmental challenges and strong degradation at the ecosystem level, including climate change, deforestation and biodiversity loss (McMichal et al. 2007; Foley et al. 2011; FAO 2017; Adelodun et al. 2021). For this reason, innovation in the production and proposal of sustainable, smart and safe solutions present the main challenge for the agri-food sector (United Nations 2015; Aydin and Parker 2018; Contini et al. 2020), and as a response to these challenges, AVCs need to integrate the sustainability approach to have a competitive advantage (Soosay et al. 2012).

Firstly, the concept of "sustainability" refers to a general system that focuses on environmental, social and economic elements and meets the needs of new generations (FAO 2014; Ingram et al. 2016; Chotia and Rao 2018; World Bank 2020). In this context, we can define a sustainable agricultural value chain as a set of successive and coordinated activities of valorization of enterprises, which produce agricultural raw materials and transform them into food products sold to final consumers and disposed of after use in a cost-effective manner throughout the process that has benefits for society and does not permanently deplete natural resources (Soosay et al. 2012; FAO 2014). In the pursuit of this goal, the strategic actors in AVCs must join efforts to implement a sustainable strategy. In this context, producers work more closely and independently of other stakeholders along the chain with the goal of producing in a more socially, economically and environmentally responsible manner (FAO 2014).

Immediately after planting, the young plantation must be directly irrigated. Regarding this topic, the irrigation of date palms is considered to be an essential element not only for strategic projects but also as a central element of the water budget for any agricultural area. In this context, in the pursuit of optimal water use, producers apply modern microirrigation methods (drips) to preserve this scarce resource (Elfeky and Elfaky 2019; Ainaim et al. 2022). In Tunisia, for example, the first farmers highlighted the problem posed by the Nefta spring, which once provided some 700 L of water per second. The water resources were quickly exhausted, posing a significant problem to the agriculture of the date "Deglet Nour". To remedy this, instead of continuing to use the original irrigation system of flooding orchards with water pumped 100 m underground and wasting water, date producers have been experimenting with new technologies such as microirrigation. In addition, fertilizer application can be integrated with drip irrigation for further cost reduction (Hema et al. 2012). On the other hand, farmers take care to wrap the ripening dates in a well-ventilated cloth to keep birds and insects away, which also avoids the use of pesticides (Bateel official website).

6.2.3.4. *Marketing of dates*

According to Food Agriculture Organization statistics in 2019, the global date production volume is approximately 9.07 million metric tons, compared to 8.4 million metric tons in 2017. We find that the Middle East and North Africa account for 90% of global production. As shown in Figure 6.1, Egypt is the world's largest producer of dates, followed by Saudi Arabia and Iran (FAO 2019).

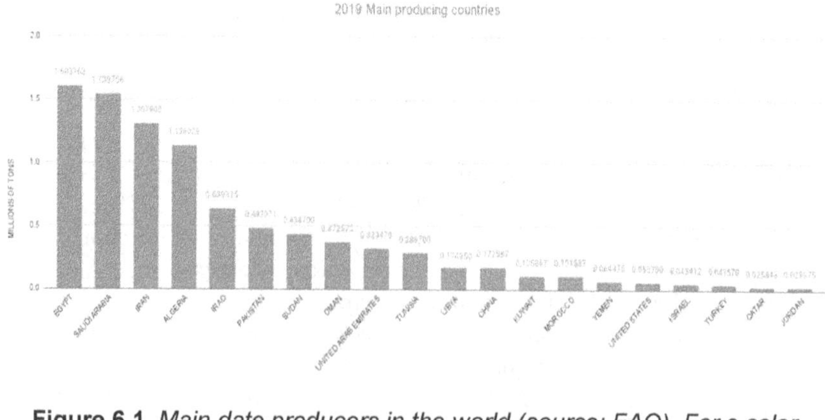

Figure 6.1. *Main date producers in the world (source: FAO). For a color version of this figure, see www.iste.co.uk/casadella/agri-innovations.zip*

These countries consume almost all their production, which is why Tunisia, despite its low production compared to other date-producing countries, occupies first place in terms of exports (Figure 6.2).

		2020 exported value (billions of USD)	**Quantity exported in 2020**	**Average unit value (USD/tonnes MT)**	**Growth p.a. value 2016–2020 (%)**	**Quantity p.a. value 2016–2020 (%)**	**Share of world exports**
	World	1,850,922	1,270,172	1,457	11	9	
1	Tunisia	298,903	117,080	2,553	6	2	16.1
2	Saudi Arabia	245,969	213,411	1,153	14	16	13.3
3	Israel	235,746	62,689	3,761	13	–2	12.7
4	United Arab Emirates	150,040	163,832	916	30	36	8.1
5	Iran	139,403	189,737	736	4	10	7.5
6	Algeria	129,191	67,483	1,914	27	22	7
7	United States	89,508	18,923	4,730	12	14	4.8
8	Iraq	78,800	176,339	447	4	3	4.3
9	Pakistan	60,264	116,047	519	–14	–12	3.3
10	The Netherlands	53,077	10,141	5,234	13	8	2.9
11	France	48,605	18,163	2,676	7	9	2.6
12	South Africa	47,463	17,793	2,668	51	54	2.6
13	Egypt	41,424	9,712	4,265	3	–30	2.2
14	Germany	39,515	10,542	3,748	19	18	2.1
15	Palestine	31,872	5,602	5,689	33	31	1.7

Figure 6.2. *Major date exporters in the world in 2020 classified by value developed from the trademap (source: GIZ 2021)*

We emphasize that there are non-date producing countries such as France and Germany, which are among the most important importers in Europe and the most important supplier or "re-exporter" of dates from Europe (Bernard 2000). Thus, each date-producing country is not systematically a producing country.

The date can be marketed all over the world as a high-value confectionery. Before its marketing, it is imperative to ensure that this fruit meets the standards of hygiene and quality. In this logic, dates go through several stages of processing and recovery to be classified according to their qualities. The dates, arrived at their maturity, are collected and then sold according to their varieties and qualities. Among the official standards of marketing and quality control of dates, we find the standard UNECE DDP-08 at the European level (2021), which classifies dates into the following three categories:

– Extra or high-quality dates: these dates have the best quality insofar as the production process is conditioned by quality standards and food safety system.

– Category 1: this category refers to dates of medium ranges that may have some slight defects that do not affect the overall quality, appearance of the product, or its conservation or presentation. This category is from supply chains that do not pay much attention to quality and food safety, hence the possibility of presenting a slight defect in the epicarp not affecting the pulp, a slight defect in shape or development and slight wrinkles.

– Category 2: these are dates that meet the minimum quality characteristics and retain the general appearance of quality, conservation and presentation. This class may include the following defects: defects of the epicarp not affecting the pulp, defects in shape or development and defects in coloring.

There are two types of dates: (1) natural dates called "fresh" (pitted, branched or bulk) and (2) conditioned dates (packaged). Packaging is the final stage of processing of the product that meets the requirements and conditions listed in the specifications. Through the following figure, we simplify the different stages of transition of dates in a packaging station (Gendre et al. 2007; APII 2017).

In Tunisia, the most exported dates abroad are the variety of "Deglet Nour". Eighty percent of this variety is exported after its treatment at the level of the conditioner-exporter link (El Moueddeb et al. 2011). In the first phase and after the receipt of dates by producers or collectors, the packers check and control their quality and classify dates into three categories of harvested products: dates branched, dates in bulk of good quality (extra or category 1) and dates of inferior quality (category 2). Then, comes the stage of fumigation based on phosphine (PH3) and/or CO_2, which aims to disinsectize dates by eliminating larvae and insect eggs before they hatch. The precise sorting is done just after fumigation, and during this stage, dates are divided into homogeneous groups according to their colors, textures, infestations, moisture, degree of maturity and sizes (Gendre et al. 2007; APII 2017).

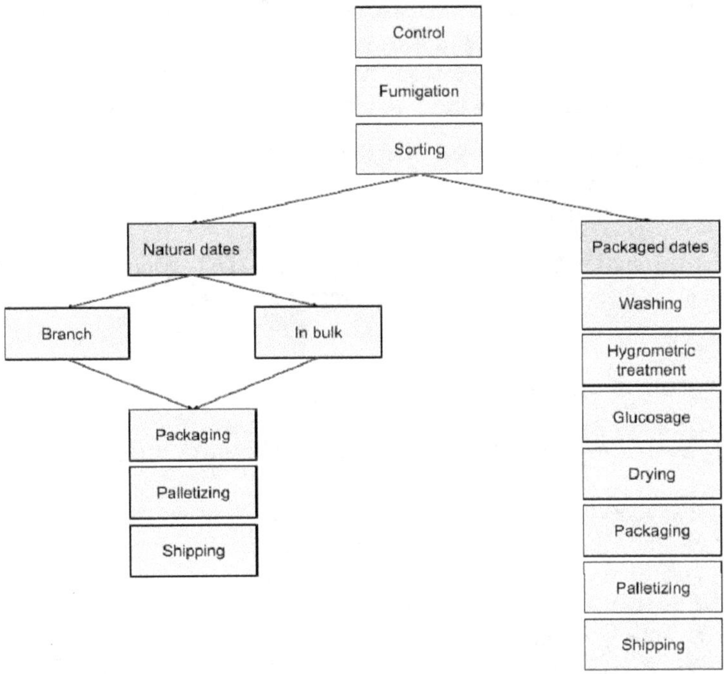

Figure 6.3. *Stage of packaging of the date (source: Agence de promotion de l'industrie et de l'innovation (APII), "Analyze de la filière des dates" 2017). For a color version of this figure, see www.iste.co.uk/casadella/agri-innovations.zip*

In the context of processing, we are witnessing a considerable valorization of dates of low marketable quality that is developing increasingly internationally. The valorization of date production is done through the proposal of countless products derived from the date; this is what generates the diversification of products on the local market, as well as on the import market (Estanove 1990; Harrak and Boujnah 2012; APII 2017; Harrak 2022). Indeed, the by-products of the date are diverse and can be grouped into three groups:

– *diversified products*: date pastes, date flours, filled dates, coated dates, and other novelties;

– *processed products*: date syrup, beverages, industrial vinegar, food vinegar, yeast, industrial or surgical alcohol;

– *products formed from waste*: date sugar, livestock feed and compost preparation.

As soon as the final product is prepared, the packer exporters move on to the next step, which is the labeling and packaging of their products in an appropriate manner according to European rules. As an example, it aims to allow consumers to make informed choices about the products they consume and their composition. To be marketed, packaging must perform two types of functions (Bachiri 2020):

– traditional functions: the packaging must fulfill the technical roles of containment and transport, ensuring protection from external aggressions, and sales through the proposal of attractive packaging for the consumer;

– modern functions: packaging must provide an economic and ecological function that is reusable and recyclable to protect the environment.

6.3. Background and research design

6.3.1. *The economic and social context of the date industry in Tunisia*

The global market for the date represented, in 2020, nearly 9 million tons per year. During the Mediterranean date forum held in 2022 (FOPRODATTE 2022), bringing together professionals of the date, an overview of the local and global situation of the date sector was presented. The south of Tunisia is endowed with oases constituting important centers of date production. They are spread mainly on the four governorates: Gabes, Gafsa, Kebili and Tozeur, and cover a total area exceeding 41,000 ha. Generally, phoeniculture activity is the main source of employment in these rural areas of southern Tunisia, in which we find approximately 10% of the total Tunisian population. However, it is rather a seasonal job for the majority of employees.

6.3.2. *Methods and data*

This chapter adopts qualitative research as a methodological approach. Two types of data sources were used: (1) semi-structured interviews with national and international experts in the date sector in Tunisia and around the world and (2) governmental data that enriched and complemented our remarks. The following box summarizes our research design.

Data collection: qualitative analysis was chosen as the research methodology for this chapter for two reasons. Indeed, our theme concerns a field that is insufficiently (or even very little) exploited in the management science literature; it is therefore necessary to construct it with more precision before considering testing it.

Data collection: the primary data source corresponds to semi-structured interviews conducted with date industry professionals in Tunisia and internationally. In total, five interviews were conducted. These lasted between 1 and 2 hours. They were recorded and then transcribed. The interview guide was structured into three main parts: (1) the current situation of the date industry in Tunisia and internationally; (2) the main strategic, economic, technical and social challenges faced by organizations in the sector; and (3) the prospects of the national sector and potential innovations. The nature of the questions asked varied, and we conducted several interviews with an average duration of 1 hour and 30 minutes, which allowed us to ask for clarification on data collected previously.

The second source of information is based on government data such as the Ministry of Agriculture in Tunisia, the CEPEX or the GiFruit, which is a central agency created following the merger of the interprofessional group of dates (GID), citrus and fruit (GIAF) and the vine (GIVI). Its mission is to regulate the markets of Tunisian agricultural products, monitor the promotion of fruit quality and contribute to the marketing and promotion of exports. Gifruit is an administrative institution under the supervision of the Ministry of Agriculture, and among the functions of this grouping is the coordination of different stakeholders, whether in the private sector or public sector at different levels, at the level of production, training of stakeholders and at the level of export, and it is managed by a board of directors.

Box 6.1. *Methods*

The profiles of the interviewees who underwent a semi-structured interview are detailed in Table 6.1.

	Activities	**Duration of the interview**
EXP1	International date consultant	1 h
EXP2	International consultant in agriculture	1 h 30 min
EXP3	Consultant in agronomy	1 h
EXP4	Expert to international institutions	1 h 30 min
EXP5	Project manager of the Tunisian professional organization Gifruit	1 h

Table 6.1. *Research interviews*

6.4. The results of the research

6.4.1. *An agricultural value chain in difficulty despite its global leadership*

The value chain of dates in Tunisia has shone for many years on the world market with its variety of Deglet Nour. However, its leadership position in terms of exports in value is threatened because of very strong international competition. Countries, formerly importers such as Morocco, are entering the same market with the variety Mejdoul. Even if the variety Deglet Nour, produced by Tunisia and Algeria, has always dominated the European market, it will face strong competition in the medium term. In recent years, it has struggled not to lose its market share to the gradual growth of the Medjoul variety, which is exported under the label Medjoul Star, produced in Morocco. In addition to its good quality, the huge growth of the Medjoul variety has been the result of a favoring of local distribution channels by the European market and by the change of export destination of Iranian and Pakistani Medjoul dates and their orientation toward the Russian and Indian markets. Morocco is therefore in an impressive competitive position thanks to its geographical proximity to the European market.

While the Maghreb threat seems important internationally, local difficulties are equally important. Indeed, the value chain of dates has important obstacles to its structuring. Firstly, the various links that make it up are not structured around a collective strategic consensus (Boughzala et al. 2019). The collectors' link seems to be in the most difficulty due to the heterogeneity of the actors and the economic and social precariousness of this professional category. On the other hand, the link dedicated to exports seems to be the most organized. This can be explained in particular by the requirement for product quality and logistics by international clients.

Moreover, despite considerable support from support agencies, the date value chain is facing significant financial and technological difficulties. The ingenious irrigation system of date palms, such as the technique of self-regulating drippers, particularly adapted to clay soils whose permeability is more compact, will no longer suffice to meet the imminent environmental problem related to water stress. Other irrigation techniques are being adopted, such as mini-diffusers, bubbler and subsurface irrigation. An important study on date palm irrigation systems (Dhaouadi et al. 2015) shows that the different current systems, despite their performance noticed in the 1980s, are no longer efficient. Comparative work has shown that the quantities of irrigation water are quickly exhausted for the four techniques studied. The environmental issue seems to be more important than ever for the agricultural sector and the local economy.

Finally, the annual harvest includes a considerable percentage of unsold and lost dates (dates not fit for consumption). The estimate is not specified with precision by the various experts, but it is evaluated, according to the answers of our interviewees, to be between 35% and 45% of the total annual production.

> Even if we do not plant any date palms in a few years, we will still have 200 thousand tons of dates for export, 100 thousand tons of losses and another 100 thousand tons of unsold dates (EXP2).

Unsold and lost dates are a major concern for the industry. This is both a business opportunity if the date is processed or if new markets are negotiated and a potential economic threat if the problem is not strategically anticipated and resolved.

6.4.2. *The guarantors of global leadership*

The difficulties associated with the date value chain in Tunisia threaten the performance of its value export business. However, these difficulties can be transformed into challenges if sector organizations overcome these various obstacles by strengthening their capacities.

6.4.2.1. *The diversification of markets and the transformation of the date into products appreciated by the current responsible consumption*

The diversification of markets is a first option for development. Some of the experts interviewed propose to produce the Medjoul variety in Tunisia to enrich its commercial offer and not to reduce its activity only to the Deglet Nour date. This development option involves planting new date palms of this variety. It is an investment in the medium term that can pay off.

Following the various rivalries on the international level, Tunisian actors have quickly sought solutions to enhance their products and stand out from competitors by offering organic dates. In this context, the Agricultural Development Group (GDA) Bijoux de l'Oasis in Tunisia perfectly illustrates the willingness of Tunisians to enhance their dates and preserve their position through the formation of sustainable strategic partnerships.

> They have partnerships with two French fair trade companies: one company that imports date sugar, and the other fresh dates. They work with Ressources bio, and Ethiquable (EXP1).

In addition, the processing of the date seems to be a strategic point for organizations in this sector. The considerable percentage of losses and unsold call decision makers to move toward a transformation of the date into various products such as date sugar, date syrup or coffee made from date pits. This measure coincides with a European consumption pattern more oriented toward responsible products, without chemical additives and are beneficial to health. Moreover, thanks to its natural benefits, especially in terms of good-quality carbohydrates and dietary fiber, dates are a major ingredient in nutritious snacks prized by European consumers. The experts also propose an "education of the Tunisian consumer" by promoting derivatives of the date, which will free people from the addiction to refined sugar, often associated with health problems. In addition, on the organizational level, startups dedicated to the transformation of the date into food by-products called "dates of mouth" or cosmetic derivatives are emerging and are experiencing considerable growth on the international market.

Finally, the organizations of this sector would benefit from internationalizing on other markets, mainly African markets with a Muslim population likely to consume dates, especially during the month of Ramadan, where consumption is multiplied.

> This implies developing trade agreements and international logistics oriented toward the countries of the South whose date markets are still very underdeveloped (EXP5).

6.4.2.2. *Certification and establishment of a label*

The competition in the Maghreb for dates is quite intense. Algerian exporters export Algerian dates softer and fleshier than Tunisian dates. They are more appreciated by the North African consumer. On the other hand, competition between Tunisian farmers offering unbeatable prices to French importers is not negligible. Moreover, with the crisis of Covid-19, consumers around the world have become selective about the origin and manner in which agricultural products are grown. This intensification of the competition within the sector has pushed the strategic actors of the GVC to focus on the certification of their dates to guarantee a higher and satisfactory quality to the final consumer. BENI GHERIB, a company which collects, packages and exports dates, mainly organic and biodynamic dates, is the first to be certified organic and biodynamic dates in Tunisia and fair trade in the world.

> They have obtained the organic certificates, biosuisse, demeter, fair trade, NOP for the USDA organic, it is the part of the production certified with our group and we have the certificate ISO 22000 and ISO 9001. Partially, they are interested in the implementation of the IFS, which is a reference frame of audits set up by the big distribution. This consists of a certification guaranteeing the safety and quality of food products. They want to integrate into the global market and find a place with other importers. This requires more requirements at the moment for the certification of processes and at the level of transformation (EXP3).

The certification for the organizations producing the date in Tunisia and the establishment of a fair trade label on a more global scale is a significant way to sustainability and viability of the sector. This policy of establishing quality is also a guarantee in the eyes of European consumers, who are more concerned about the quality and provenance of food products. The *Made in Tunisia*, embodied by the date Deglet Nour, must be renewed. The competitive game is no longer based solely on the characteristics of the date as a fruit with many virtues but on the ability of organizations that produce, package and market it. It is for Tunisian producers to enhance their organization by strengthening structural, managerial and strategic capacity, then the government to strengthen the sector by establishing a label.

6.4.3. *Challenges of sustainability*

6.4.3.1. *The development of technological solutions*

Long used in the industrial sector, technological innovation also seems to be inviting itself to the agricultural sector. In this regard, Agriculture 4.0 is restructuring organizations from upstream to downstream of the value chain. It can be used by producers for more efficient irrigation systems, as well as at all levels up to the final consumer for a better traceability of products. To date, few producers are using blockchain. Experts point out the technological difficulties related to the entire sector. The latter needs to be modernized, not only at the level of exports that are proceeding to a more qualitative industrialization than the rest of the sector. The exporters have extensive experience in the field, and their oligopoly stifles the possibilities of structural and technological value to other links. However, some Tunisian startups such as Seabex are experiencing significant growth. This Tunisian startup invents new technological processes to improve productivity, yield and use of water resources to cope with climate change. It has succeeded in establishing various agricultural partnerships in Europe.

6.4.3.2. *Social issues and development of the social solidarity economy*

The agricultural sector gathers 80% of the structures dedicated to the social and solidarity economy (SSE) in Tunisia. It is represented by emblematic structures of the SSE such as agricultural cooperatives in the form of GDA (*Groupement de Développement Agricole*) of SMSA (*Sociétés Mutuelles de Services Agricoles*) or more recently to solidarity enterprises. The objective of these organizations is to pool the resources of small farmers. The contribution of the SSE in this sector is not negligible. This form of economy makes it possible to guarantee decent work, particularly for women in rural areas, where the unemployment rate is one of the highest in the country. Moreover, SSE makes it possible to find sustainable employment solutions in a precarious sector, where the informal market dominates.

According to experts, the SSE can be a way toward the development of the date industry in Tunisia. The social experiences of Jemna or the production of mosquito nets protecting date palms by rural women in Nefta during the Covid-19 pandemic testify to the success of this economic and social path.

> SSE will certainly be the economy to take pressure off the sector, at all levels (EXP4).

6.5. Conclusion

This chapter analyzes the date industry in Tunisia by highlighting its potential value added, whose economic and social benefits are considerable for the national economy. The results raised the various issues that arise for the actors of this sector as well as the economic, social and technological impacts that are imposed on organizations.

For decades, Tunisia has been able to maintain its leadership in terms of exports in value because of a satisfactory level of quality and contractual links established with European partners since the 1940s. Its export logic remains essentially commercial. The price battle between local and international competitors undermines business opportunities and the possibility of establishing lasting partnerships with international customers. Additionally, in 2022, Tunisia lost its leadership position in terms of export values to Saudi Arabia. The growth and success of some varieties of dates also threaten the date Deglet Nour. Three points can be developed as recommendations for organizations involved in the sector. (1) The requirement of total quality remains a major axis for the strategic maintenance of the date sector in Tunisia. It is essential to activate the certifications and labels to reassure consumers increasingly demanding but also to give legitimacy to Tunisian companies wishing to be more competitive on the international market. Additionally, the organizations of the date sector must adapt to the fifth food transition marked by a sophisticated demand for nutritional qualities, environmental and social. (2) Innovation of products and processes seem to be required to organizations involved in the sector. Environmental threats, especially water stress, are pushing the actors in the sector to collaborate by developing interorganizational relationships (Uzunidis 2022) to pool resources and skills. (3) The development of the sector also requires strong territorial anchoring by supporting local initiatives and ensuring the balance of wealth, a condition that has become essential, especially by European customers of the labeled organic market.

Thus, technological and social innovation must be at the heart of organizational strategies. They will allow the sector to be modernized, sustainable commitments to be developed and the requirements of the global market to be met. This requires close collaboration and alliances between organizations in Tunisia, the Maghreb and the Mediterranean countries. These interorganizational relationships seem essential for the structuring and development of the date sector in Tunisia.

6.6. References

Adelodun, B., Kareem, K.Y., Kumar, P., Kumar, V., Choi, K.S., Yadav, K.K., Khan, N.A. (2021). Understanding the impacts of the COVID-19 pandemic on sustainable agri-food system and agroecosystem decarbonization nexus: A review. *Journal of Cleaner Production*, 318, 128451.

van Agtmael, A. (2007). *The Emerging Markets Century: How a New Breed of World-class Companies is Overtaking the World.* Simon and Schuster, New York.

Al-Alawi, R.A., Al-Mashiqri, J.H., Al-Nadabi, J.S., Ah-Shihi, B.I., Baqi, Y. (2017). Date palm tree (*Phoenix dactylifera* L.): Natural products and therapeutic options. *Frontiers in Plant Science*, 8, 845.

Al-Farsi, M.A. and Lee, C.Y. (2012). The functional values of dates. In *Dates: Production, Processing, Food, and Medicinal Values*, Manickavasagan, A., Mohamed Essa, M., Sukumar, E. (eds). CRC Press, Boca Raton, FL.

Al-Yahyai, R. and Manickavasagan, A. (2012). An overview of date palm production. In *Dates: Production, Processing, Food, and Medicinal Values*, Manickavasagan, A., Mohamed Essa, M., Sukumar, E. (eds). CRC Press, Boca Raton, FL.

Alanakis, C.M., Rizou, M., Aldawoud, T.M.S., Ucak, I., Rowan, N.J. (2021). Innovations and technology disruptions in the food sector within the COVID-19 pandemic and post-lockdown era. *Trends in Food Science & Technology*, 110, 193–200.

Ali, I., Satie, S., Thai, V. (2021). Adopting industry 4.0 technologies in agri-food supply chains: An exploratory investigation of drivers and barriers. In *Supply Chain 4.0: Improving Supply Chains with Analytics and Industry 4.0 Technologies*, Aktas, E., Bourlakis, M., Minis, I., Zeimpekis, V. (eds). Kogan Page Press, London.

Allani, H. (2019). Comportement d'adoption de référentiels de sécurité sanitaire des aliments chez les exportateurs de dattes en Tunisie et impact sur la performance à l'export. PhD Thesis, National Agronomic Institute of Tunisia, Tunis.

Alnaim, M.A., Mohamed, M.S., Mohammed, M., Munir, M. (2022). Effects of automated irrigation systems and water regimes on soil properties, water productivity, yield and fruit quality of date palm. *Agriculture*, 12(3), 343.

Amiel, F., Laurans, Y., Muller, A. (2019). Les chaînes de valeur agricole au défi de la biodiversité : l'exemple du cacao-chocolat. Report, IDDRI, Paris.

APII (2017). L'Agence de promotion de l'industrie et de l'innovation, analyse de la filière des dattes. Report, APII, Tunis.

Arora, C., Kamat, A., Shanker, S., Barve, A. (2022). Integrating agriculture and industry 4.0 under "agri-food 4.0" to analyze suitable technologies to overcome agronomical barriers. *British Food Journal*, 124(7), 2061–2095.

Aydin, A. and Parker, R.P. (2018). Innovation and technology diffusion in competitive supply chains. *European Journal of Operational Research*, 265(3), 1102–1114.

Bachiri, F.Z. (2020). Conception de la chaîne logistique de production des dattes en Algérie. PhD Thesis, École Supérieure des Sciences Appliquées d'Alger (ESSA), Algiers.

Bernard, O. (2000). Étude des principaux marchés européens de la datte et du potentiel commercial des variétés non traditionnelles. Étude réalisée pour le Groupe des produits horticoles. Service des matières premières et des produits tropicaux et horticoles. Division des produits et du commerce international. FAO, Rome.

Boughzala, Y., Loukil, F., Rouached, L. (2019). Partnership and capacity building in the date sector in Tunisia: The contribution of support organizations. In *Conference: Spring of Innovation, Agro-Innovation, Food Quality and Safety*. Athens.

Cagliano, R., Worley, C.G., Caniato, F.F.A. (eds) (2016). The challenge of sustainable innovation in agri-food supply chains. In *Organizing Supply Chain Processes for Sustainable Innovation in the Agri-Food Industry (Organizing for Sustainable Effectiveness)*. Emerald Group Publishing Limited, Bingley.

Carimentrand, A. (2012). Filière. In *Dictionnaire du commerce équitable*, Blanchet, V. (ed.). Éditions Quæ, Versailles.

Carrasco, E., Morales-Rueda, A., Garcia-Gimeno, R.M. (2012). Cross-contamination and recontamination by Salmonella in foods: A review. *Food Research International*, 45(2), 545–556.

Caso, D., Guidetti, M., Capasso, M., Cavazza, N. (2022). Finally, the chance to eat healthily: Longitudinal study about food consumption during and after the first COVID-19 lockdown in Italy. *Food Quality and Preference*, 95, 8.

Cavusgil, S.T. and Ghauri, P.N. (1990). *Doing Business in Developing Countries*. Routledge, London.

CEE-ONU (2010). Normes CEE-ONU DF-08 concernant la commercialisation et le contrôle des dattes entières. Report, CEE-ONU.

Chotia, V. and Rao, N.V.M. (2017). Investigating the interlinkages between infrastructure development, poverty and rural–urban income inequality: Evidence from BRICS nations. *Studies in Economics and Finance*, 34(4), 466–484.

Choudhury, D.P. (2018). *Sustainability Management*. Zorba Books, Gurugram.

Contini, C., Marotta, G., Torquati, B. (2020). Multi-actor approaches to implement cooperative strategies and value chains based on sustainability. *Agricultural and Food Economics*, 8(1), 1–4.

De Marchi, V., Giuliani, E., Rabellotti, R. (2018). Global value chains offer developing countries learning and innovation opportunities? *The European Journal of Development Research*, 30(3), 389–407.

Deloitte, N. (2013). The food value chain: A challenge for the next century [Online]. Available at: https://www2.deloitte.com/content/dam/Deloitte/global/Documents/Consumer-Business/dttl_cb_Food%20Value%20Chain_Global%20POV.pdf [Accessed 24 April 2023].

Devaux, A., Torero, M., Donovan, J., Horton, D. (2018). Agricultural innovation and inclusive value-chain development: A review. *Journal of Agribusiness in Developing and Emerging Economies*, 8(1), 99–123.

Dhaouadi, L., Maachia, B.E.N.S., Mkademi, C., Oussama, M., Daghari, H. (2015). Étude comparative des techniques d'irrigations sous palmier dattier dans les oasis de Deguache du Sud Tunisien. *Journal of New Sciences, Agriculture and Biotechnology*, 18(3), 658–667.

El Hadrami, A. and Al-Khayri, J.M. (2021). Socioeconomic and traditional importance of date palm. *Emirates Journal of Food and Agriculture*, 24(5), 371.

El Moueddeb, K., Mokhtar, B., Issam, B. (2011). Désinsectisation des dattes deglet Ennour par traitement thermique à l'aide de l'air chauffé et recirculé. *MHA (Sousse)*, 23(66), 62–69.

Elfeky, A. and Elfaky, J.A. (2019). A review: Date palm irrigation methods and water resources in the Kingdom of Saudi Arabia. *Journal of Engineering Research and Reports*, 9, 1–11.

Espiard, E. (2002). *Introduction à la transformation industrielle des fruits*. Éditions Tec & Doc, Cachan.

Estanove, P. (1990). Note technique : valorisation de la datte. Report, CIHEAM-IAMM.

FAO (2014). *Developing Sustainable Food Value Chains: Guiding Principles*. FAO, Rome.

FAO (2017). *The Future of Food and Agriculture: Trends and Challenges*. FAO, Rome.

Fernandez-Stark, K., Coutu, V., Bamber, P. (2019). *Industry 4.0 in Developing Countries: The Mine of the Future and the Role of Women*. Duke University Global Value Chains, Durham, NC.

Foley, J.A., Ramankutty, N., Brauman, K.A., Cassidy, E.S., Gerber, J.S., Johnston, M., Zaks, D.P. (2011). Solutions for a cultivated planet. *Nature*, 478(7369), 337–342.

FOPRODATTE (2022). Forum méditerranéen des professionnels de la datte organisé par l'équipe tunisienne du projet ARIMNET2 financé par l'Agence Nationale de la Recherche en France et le ministère tunisien de l'enseignement supérieur, Tunis.

Gaveau, D.L., Pirard, R., Salim, M.A., Tonoto, P., Yaen, H., Parks, S.A., Carmenta, R. (2017). Overlapping land claims limit the use of satellites to monitor no-deforestation commitments and no-burning compliance. *Conservation Letters*, 10(2), 257–264.

Gendre, L., Le Gal, P.Y., Rhouma, A. (2007). Organisation de la chaîne d'approvisionnement de la datte tunisienne. Étude dans le cadre du Projet Sirma : économies d'eau en systèmes irrigués au Maghreb, Cirad, et Centre de Recherche Régional pour l'Agriculture Oasienne (CRRAO). Student, Projet Sirma.

Gerefi, G. (1999). International trade and industrial upgrading in the apparel commodity chain. *Journal of International Economics*, 48(1), 37–70.

Gerefi, G. and Fernandez-Stark, K. (2016). *Global Value Chain Analysis: A Primer*, 2nd edition. Duke University, Durham, NC.

Gerefi, G., Humphrey, J., Sturgeon, T. (2005). The governance of global value chains. *Review of International Political Economy*, 12(1), 78–104.

Gesellschaft fur internale zusammenarbeit (2021). *Exporter des dattes en Europe : perspectives du marché pour votre produit*. Gesellschaft fur internale zusammenarbeit, Berlin.

Glover, J.L., Champion, D., Daniels, K.J., Dainty, A.J. (2014). An institutional theory perspective on sustainable practices across the dairy supply chain. *International Journal of Production Economics*, 152, 102–111.

Harrak, H. (2022). Main achievements of research in improving and developing Moroccan date postharvest valorization: A review. *AFRIMED AJ-Al Awamia*, 134, 1–29.

Harrak, H., Boujnah, M., Arrak, H., Boujnah, M. (2012). *Valorisation technologique des dattes au Maroc*. INRA Éditions, Rabat.

Hema, N., Krishna, K., Maringanti, H.B. (2012). Site specific automated drip irrigation for palm trees using wireless sensor network. In *Information Technology, Automation and Precision Farming, International Conference of Agricultural Engineering-CIGR-AgEng: Agriculture and Engineering for a Healthier Life*, 8–12 July. Valencia.

Hidayati, D.R., Garnevska, E., Childerhouse, P. (2021a). Sustainable agrifood value chain – Transformation in developing countries. *Sustainability*, 13(22), 14.

Hidayati, D.R., Garvevska, E., Childerhouse, P. (2021b). Transforming developing countries agrifood value chains. *International Journal on Food System Dynamics*, 12(4), 358–374.

Humphrey, J. and Memedovic, O. (2006). Global value chains in the agrifood sector. Working Paper, United Nations Industrial Development Organization, Vienna.

Ingram, V.J., Judge, L.O., Luskova, M., Van Berkum, S., Van Den Berg, J. (2016). Upscaling sustainability initiatives in international commodity chains: Examples from cocoa, coffee and soy value chains in The Netherlands. Statutory Research Tasks Unit for Nature and the Environment, No. 67.

Jones, L., Demirkaya, M., Bethmann, E. (2019). Global value chain analysis: Concepts and approaches. *Journal of International Commerce and Economics*, 29.

Kamath, R. (2018). Food traceability on blockchain: Walmart's pork and mango pilots with IBM. *The Journal of the British Blockchain Association*, 1(1), 3712.

Khanna, T. and Palepu, K.G. (2010). *Winning in Emerging Markets: A Road Map for Strategy and Execution*. Harvard Business Press, Brighton.

Korpela, K., Hallikas, J., Dahlberg, T. (2017). Digital supply chain transformation toward blockchain integration. In *Proceedings of the 50th Hawaii International Conference on System Sciences*. Hawaii.

Lancon, F., Temple, L., Biénabe, E. (2017). The concept of filière or value chain: An analytical framework for development policies and strategies. In *Sustainable Development and Tropical Agri-chains*, Biénabe, E., Rival, A., Loeillet, D. (eds). Springer, Dordrecht.

Lescuyer, G. (2020). Towards a hybridization of the cocoa sector governance in Cameroon to meet economic and environmental sustainability. Report, CGIAR Research Program on Forests, Trees and Agroforestry (FTA).

Lezoche, M., Hernandez, J.E., Diaz, M.D.M., Panetto, H., Kacprzyk, J. (2020). Agri-food 4.0: A survey of the supply chains and technologies for the future agriculture. *Computers in Industry*, 117, 103187.

Lin, X., Chang, S.C., Chou, T.H., Chen, S.C., Ruangkanjanases, A. (2021). Consumers' intention to adopt blockchain food traceability technology towards organic food products. *International Journal of Environmental Research and Public Health*, 18(3), 912.

Liu, Y., Eckert, C., Yannou-Le Bris, G., Petit, G. (2019). A fuzzy decision tool to evaluate the sustainable performance of suppliers in an agrifood value chain. *Computers Industrial Engineering*, 127, 196–212.

Manfroy, S. (2021). Analyse de la chaîne de valeur café en République démocratique du Congo selon la méthode VCA4D : cas de la Réserve de Biosphère de Luki. Master's Thesis, Université catholique de Louvain.

McMichal, A.J., Powles, J.W., Butler, C.D., Uauy, R. (2007). Food, livestock production, energy, climate change, and health. *Lancet*, 370, 1253–1263.

Mergenthaler, M., Weinberger, K., Qaim, M. (2009). Quality assurance programs and access to international markets: The case of horticultural processors in Vietnam. *Supply Chain Management*, 14(5), 359–368.

Mwanbi, M.M., Oduol, J., Mshenga, P., Saidi, M. (2016). Does contract farming improve smallholder income? The case of avocado farmers in Kenya. *Journal of Agribusiness in Developing and Emerging Economies*, 6(1), 2–20.

Nestlé (2019). Nestlé lays out action plan to help end deforestation and restore forests in the cocoa supply chain. Report, Nestlé, Vevey.

Olaimat, A.N., Shahbaz, H.M., Fatima, N., Minur, S., Holley, R.A. (2020). Food safety during and after the era of COVID-19 pandemic. *Frontiers in Microbiology*, 11, 1854.

Pananond, P. (2022). Crises, emerging market firms, and global value chain resilience. In *International Business in Times of Crisis: Tribute Volume to Geoffrey Jones*, van Tulder, R., Verbeke, A., Piscitello, L., Puck, J. (eds). Emerald Publishing Limited, Bingley.

Pappa, I., Illiopoulos, C., Massouras, T. (2019). On sustainability of a dairy sector in crisis. *International Journal on Food System Dynamics*, 10(2), 130–150.

Ranomenjanahary, S.B. (2018). Accroissement de revenus des ménages par l'agriculture à travers l'organisation des producteurs et l'approche SILC dans la commune d'anivoranoest (district brickaville). Master's Thesis, University of Toamasina.

Schoneveld, G.C., van der Haar, S., Ekowati, D., Andrianto, A., Korarudin, H., Okarda, B., Pacheco, P. (2009). Certification, good agricultural practice and smallholder heterogeneity: Differentiated pathways for resolving compliance gaps in the Indonesian oil palm sector. *Global Environmental Change*, 57, 101933.

Shaffer, J.D. (1973). On the concept of subsector studies. *American Journal of Agricultural Economics*, 55(2), 333–335.

Sharma, R. (2022). Industry 4.0 technologies in agri-food supply chains: Key performance indicators. In *Agri-Food 4.0*, volume 27, Mor, R.S., Kumar, D., Singh, A. (eds). Emerald Publishing Limited, Bingley.

Skalkos, D. (2022). Innovative agrifood supply chain in the post-COVID 19 era. *Sustainability*, 14(9), 53–59.

Soosay, C., Fearne, A., Dent, B. (2012). Sustainable value chain analysis – A case study of Oxford Landing from "vine to dine". *Supply Chain Management: An International Journal*, 17(1), 68–77.

Taglioni, D. and Winkler, D. (2016). *Making Global Value Chains Work for Development*. World Bank Publications, Washington D.C.

Tengberg, M. (2009). Cultures et utilisations du palmier dattier au Moyen-Orient ancien. Données archéobotaniques. *Cahier des thèmes transversaux ArScAn*, 9, 237–242.

Tian, F. (2016). An agri-food supply chain traceability system for China based on RFID blockchain technology. In *2016 13th International Conference on Service Systems and Service Management (ICSSSM)*. Kunming.

Time Is (2019). Dates farming [Online]. Available at: https://www.techno-preneur.net/technology/project-profiles/food/Farming.htm.

Trepant, I. (2008). Pays émergents et nouvel équilibre des forces. *Courrier hebdomadaire du CRISP*, 19911992, 6–54.

Trienekens, J. and van Dijk, M.P. (2011). *Linking Local Producers from Developing Countries to International Markets*. Amsterdam University Press.

United Nations (2015). Transforming our world: The 2030 agenda for sustainable development. United Nations, Department of Economic and Social Affairs, New York.

Uzunidis, D. (2022). Aperçu de la systémique des filières. Spring of Innovation RNI, Paris.

WCED (World Commission on Environment and Development) (1987). World commission on environment and development: Our common future. Oxford University Press.

World Bank (2020). World development report 2020: Trading for development in the age of global value chains. World Bank, Washington, D.C.

Zaid, D.A. and De Wet, P.F. (1999). Date palm propagation. In *FAO Plant Production and Protection Papers*. Food and Agriculture Organization of the United Nations, Geneva.

7

Technology, Innovation and Sustainability of the Soybean Chain: Study of the Cameroonian Cotton Front Facing Environmental Challenges

7.1. Introduction

The volatility of agricultural commodity prices in international markets (cotton, cocoa, coffee, etc.)[1] and its consequences on national food production, as well as the recurrence of food crises in many African countries over the past 15 years, have repositioned agricultural issues on the agendas of governments of developing countries, developed countries and international institutions. Alongside the commitments made by the states[2] at the World Food Summit (WFS) in 1996, as well as at numerous global meetings (G8 Agriculture Summit in 2009, G20 Agriculture Summit in 2011, etc.), international and national responses to global food crises

Chapter written by Eric Joël FOFIRI NZOSSIÉ, Darryl Neil NITCHEU WAKPONOU and Christophe BRING.

1 For example, Yankou (2014), assessing the rates of change of certain commodities, notes that with an average variability of 25.8% over the period 2003–2012 according to the instability index calculated by the United Nations Trade and Development Organization (UNCTAD), cotton prices are among the most volatile, far ahead of those of wheat (18.2%) or coffee (17.3%).

2 These commitments were to increase the budgetary share of agriculture by at least 10%.

have mainly mobilized two agronomic trajectories: agricultural intensification[3] and crop diversification to increase national food supplies and improve farmers' income.

Soybean (*Glycine max* (*L.*) *Merr.*) represents a good example of the realization of this second trajectory in the Sudano-Sahelian region of Cameroon. Over the past 10 years, its spatial and production dynamics have led to the relativization and reconsideration of many pessimistic, even alarmist, judgments about the capacity of African agriculture to meet the multiple challenges of increasing and diversifying agricultural supply, improving farmers' incomes, and supplying companies in the agri-food sector. These judgments have generally been based on the weakness of agricultural growth to reduce rural poverty, as well as the low representation of agriculture in international trade, the impact of price drops and fluctuations of the main export products on producers who are gradually integrated into a market economy and productivity logic (Griffon 2003). For this author, for example, there was nothing in these trends to suggest that African agriculture was preparing to face economic, social and ecological challenges, the magnitude of which we know to be unprecedented in history, thus positioning agriculture in sub-Saharan Africa in a unique situation compared to other continents.

Experimentation and support for the dissemination and development of soybeans in the Sudano-Sahelian region of Cameroon was a strategic option for the *Société de Développement du Coton* (SODECOTON), with the objective of creating added value in the face of the volatility of cotton prices in international markets. Diversification thus responded to the need to revitalize a society polarized by the cotton industry (support for contractual production by farmers, export of fiber and production of vegetable oil from cotton grain). Soybeans will be integrated into the production process for vegetable oil and cake at SODECOTON, thus marking the company's policy of technological innovation to mitigate the effects of the cotton crisis.

Since 2008, the Sudano-Sahelian region, which includes the Cameroonian cotton basin, has experienced a dynamic in soybean production that has led to profound changes in local agri-systems and has contributed to the (re)configuration of new market relations between producers, traders, distribution intermediaries and agri-food companies. According to available official statistics, regional production increased from 41 tons in 2007 to nearly 14,000 tons in 2017, while Cameroon's total production for the same period was 7,801 and 18,886 tons, respectively[4]. This evolution has followed the global trend marked by significant growth for at least 25 years (123% between 1996 and 2004), particularly in Latin America (Arvor 2009; WWF 2014). Since 2010, soybeans have been ranked as the second most

3 Intensification in the sense of increasing yields through better access to synthetic inputs and irrigation.

4 FAOSTAT, see: http://www.fao.org/faostat/fr/#data/QC [Accessed June 23, 2020].

important legume crop in the cotton basin after groundnuts, followed by cowpeas and voandzou. Since 2018, it has been classified as a major crop in the nomenclature of Sudano-Sahelian food crops.

By 2019, nearly 60% of cotton producers had either abandoned cotton cultivation, diversified into food crops or substituted less profitable crops that are poorly supported by public and private agricultural development actors (e.g. *Vigna Unguiculata L. walp* – niébé). The dominant trend of establishing monoculture plots is leading to the reconfiguration of agrosystems and changes in crop rotation. In a growing number of family farms, cotton has been disregarded as the main crop in favor of soybeans.

Most of the region's production feeds a growing demand from the domestic agri-food sector, but also from across the border in Nigeria[5]. However, due to porous land and river borders, the proportion of production exported to Nigeria remains unknown[6]. According to macroeconomic data, Cameroon imports an average of 20,000 tons of soybeans per year for a value of nearly 10 billion CFA francs[7], as well as GMO soybean meal for a value of 14 billion[8]. Hence, there is the challenge of increasing national supply to meet agro-industrial demand. The rapid increase in cultivated areas from 6,705 ha in 2008 to 15,020 ha in 2018 (FAOSTAT) is indicative of the enthusiasm of farmers for this crop, which receives very little support from the government. Its rapid development in the cotton basin, where the income of more than 80% of agricultural workers has historically depended on the sale of cotton, offers social science research a new opportunity to examine new development dynamics in an area that has long been polarized by cotton research.

These plural dynamics are of a socioprofessional (farmers' clear desire to set up a structured link through cooperative movements), economic (construction of a value chain), technological (diversification of the product offered to consumers) and environmental nature. Although still little explored, these dynamics nevertheless raise issues of environmental preservation in a space with a fragile ecology. This last aspect is the main theme of this chapter.

---

5 Domestic consumption in the Sudano-Sahelian region remains marginal, as soybeans have not been a part of people's dietary habits until the last ten years. Soy has become more integrated into infant nutrition as a solution for reducing malnutrition, under the impetus of public health services and development actors (NGOs, United Nations agencies).

6 Since the colonial administration, the Sudano-Sahelian region has been subject to a ban on the export of food grains (cereals and legumes) to protect the population from the risk of food insecurity. This measure has been maintained and even reinforced by the Cameroonian administration. The export of soybeans, like other crops, therefore follows informal channels that are difficult to quantify.

7 15,244,900 Euros.

8 21,342,900 Euros.

The above context gives rise to the following research question: in what way do the agricultural practices that underlie the production dynamics of soybeans constitute a vector for accelerating environmental degradation in a Sudano-Sahelian area whose natural capital is increasingly threatened by the effects of climate change? The objective of this research is therefore to analyze the environmental consequences inherent in these practices. The research is based on the premise that the willingness of producers to seize the opportunity offered by the demand of national agribusiness companies and the transborder soybean market is polarized by the strategy of extending the cultivation front, which exacerbates the clearing of vegetation cover weakened by the drying of the environment and leads to the massive use of chemical substances to ensure crop protection.

7.2. State of knowledge on the subject

Soybean is now one of the major and important sources of protein (40–42%) and vegetable oil (18–22%) used in human food (Mehmet 2008; Baboy et al. 2015). However, as with many other crops, its yield is limited in sub-Saharan Africa by several factors, including poor cultural practices, inadequate land selection for its cultivation, soil fertility degradation, and climate change (Kananji et al. 2013). The orientation of this research highlights three major issues that have thus far structured the axes of knowledge production: those related to socioeconomic diversification in rural Africa, those inherent to the polarization of the productive dynamics of soybean by the agribusiness industry in Latin America and sub-Saharan Africa, and finally those related to energy diversification and the fight against climate change (Fofiri et al. 2020).

7.2.1. *Soybean in the socioeconomic diversification of rural Africa*

In the early 1970s, the IITA (International Institute of Tropical Agriculture) initiated its program to improve African soybean lines based on the crossing of high-productivity lines of Asian origin with TGx (Tropical Glycine Cross) lines. This work will allow Nigeria, for example, to boost soybean production from 75,000 tons in 1980 (yield below 300 tons/ha) to 758,033 tons in 2018 (average of 1 ton/ha) (FAOSTAT[9]). In Africa, this crop is now considered as one of the solutions that can improve the quality of food for the population, but also contribute to the diversification of farmers' income. However, the agronomic challenges for its establishment remain significant, with efforts aimed at improving yields (1.2 tons/ha on average), seed viability, fertilization and soybean inoculation.

9 FAOSTAT, see: http://www.fao.org/faostat/fr/#data/QC [Accessed June 7, 2020].

Since the 1980s, soybeans have been integrated into the agrosystems of many sub-Saharan African countries to ensure food self-sufficiency, accompanied by incentives. The lack of processing opportunities in many countries led to its abandonment by farmers. In Cameroon, it has been popularized for approximately 15 years, but national demand has been quick to take off (agro-industries producing vegetable oil, infant flour and dairy products, animal feed). The crop, which was more common in the West Cameroon region in the 1990s, was popularized in the cotton basin in the 2000s. However, this South–North spatial dynamic remains poorly documented scientifically, apart from scattered statistics on production.

7.2.2. *The agribusiness industry as a lever for strengthening the use of science and technology (S&T) policies and accelerating soybean cultivation in sub-Saharan Africa*

The agri-food industry is one of the industrial sectors that has undergone remarkable development over the last 20 years, marked by an extraordinary diversification of the supply of processed products based on cereals, legumes and oilseeds. Global statistics on meat consumption show, for example, that nearly three-fourths of the soy produced in the world is used to feed animals, particularly poultry and pork. Between 1967 and 2007, pork production increased by 294%, eggs by 353% and poultry meat by 711%. Over the same period, the relative cost of these products decreased. As the world's largest source of animal feed, soybeans have become an indispensable part of the intensive agricultural model (WWF 2014). These grains make up the bulk of dairy products, oils and flours. The agribusiness industry has thus become a lever for polarizing production dynamics in sub-Saharan Africa through national companies, but also under the impetus of global demand that is increasingly oriented toward organic products. This second axis needs to be further documented with regard to technological innovations that result in a wide diversification of the product offer (beverages, dairy products, growth foods, etc.). However, this diversification would reinforce the use of science and technique (S&T) policies for public policies in developing economies, which are known to be scarce (Casadella and Tahi 2021). In these economies, technology policies are more related to the commercialization or nonacademic valorization of scientific knowledge (Casadella and Tahi 2021) in a context of low transfer of academic research results to users (Fofiri et al. 2015).

7.2.3. *S&T policies for the valorization of soybean in energy diversification and the fight against climate change*

The last 20 years have seen an intense development of biofuel production and use worldwide. From 1996 to 2006, the tonnage increased from approximately

16 million tons to nearly 46 million (Dronne et al. 2007). This recent expansion has strengthened the relationship between agricultural (grains, oilseeds, and pulses) and energy markets, a situation fostered by national and regional policies. In 2019, the European Union (EU) authorized the import of American soybeans for biofuel production until 2021. Available statistics show that 75% of the EU soybean market share is covered by US production. However, interest in soybeans as a fuel source is contributing to their growth in countries such as Argentina, whose soy-based agrofuel production surpassed that of Brazil in 2011. Agrofuels will therefore remain a driver of soybean demand for a long time to come, and analysts are predicting the likelihood of strong production growth by 2025 (Hart Energy 2013).

From an environmental point of view, soybeans are able to capture nitrogen from the air for their own growth and the enrichment of their seeds in proteins, but also to fix it in the soil. Thus, its cultivation contributes to the fertilization of the soil, which would limit the prospects of intensifying the use of chemical inputs and its corollary on the soil and water resources. Greenhouse gas emissions related to seed production would be significantly reduced by four to five times compared to corn, for example (Schneider and Huyghe 2015). However, on the other hand, the extension of cultivated areas contributes significantly to deforestation and the degradation of vegetation cover (Le Tourneau 2004; Guéneau 2018). Thus, in general, the impact of agricultural development on the environment is increasingly underpinning research dynamics, following the example of the research initiative funded by the European Union through the "*Value Chain Analysis For Development*" (VCA4D) project from 2016 to 2019, involving 24 countries in sub-Saharan Africa, Central America and Southeast Asia[10]. Such initiatives remain limited, however, and therefore open the way for individual and spatially reduced research dynamics to consider large-scale comparative perspectives. Nevertheless, they have the merit of taking into account the recent mutations that affect African territories.

The growing interest in biofuels, especially in industrial countries, is a vector for the development of S&T policies, which are marked by the production of so-called second-generation biofuels that allow the conversion of all biomass, pushing back the limits of first-generation biofuels. These developments are supported by technological innovations that complete the range of renewable energies (Gurtler 2008). However, these technological challenges are still limited in sub-Saharan Africa.

---

10 Agrinature, *Value Chain Analysis for Development* (VCA4D), see: https://europa.eu/capacity4dev/value-chain-analysis-for-development-vca4d- (accessed June 23, 2020).

7.3. Theoretical framework of the study

The study proposes a crossing of two complementary theoretical approaches to try to grasp the environmental stakes induced by the increase in soybean production: the theory of spatial diffusion and the theory of population-environment relations. The interest in spatial diffusion theory was revived in contemporary geographical studies by Hàgerstrand (1952, 1957) with contributions from Saint-Julien (1985) and Gould (1992). These pioneering works underline the importance of temporality and spatiality in any diffusion process. Any diffusion requires contact between transmitters and receivers. These contacts imply spatial contiguity or connectedness (Dumolard 1999; Maudet 2011). The theory is appropriate for the study of processes that involve the movement of materials, products, people, practices, or ideas in the aggregate. It has three main stages: the initial stage of appearance and progressive growth of the observed phenomenon, the intermediate stage of accelerated growth, and the final stage of saturation and decay.

In the social sciences, the notion of diffusion is associated with that of innovation, from which it is inseparable. The trajectories of soybean introduction and extension are part of the dynamics of innovation in the sense of Schumpeter (1934). This innovation, which is both technical (varietal development, for example) and organizational (structuring the value chain), implies the combination of new things that propagate in an environment, generating irreversibilities in the evolution of this environment (varietal selection, technical itineraries, choice of soils or modification of crop rotation, etc.). Schematically, the application of this theory to soybean cultivation allows us to situate its current evolution in the second intermediate and accelerating stage of growth in a double productive and spatial dimension. This growth is marked by the continued clearing of new land, the availability of which in grazing areas and on the margins of defended areas is likely to continue in the absence of a real public policy to reverse the trend. This reality thus postpones the stage of saturation and decrease in land clearing in the current context of absence and/or failure of support for actors by public authorities, poor control of technical itineraries and extensification of production.

The theory of diffusion extends that of the population-environment relationship considered by several authors to be difficult to understand scientifically complex relationships (Sandron 2005), and even uncertain ones (Ouharon 2012). Indeed, the relationship between humans and the environment has founded the construction of the disciplinary base in geography, supported by the Ratzelian determinism that gives a preponderant place to the Natural Environment in the analysis and explanation of human behavior and societies. The history of the geographic discipline thus brings the definition of determinism back to the contributions of certain German geographers of the 18th century, founders of the German school of geography, notably Carl Ritter (1779–1859) and Friedrich Ratzel (1844–1904). The theory of the population–

environment relationship is, however, built around two visions derived from economics and biology (Gendreau et al. 1996). The choice of this theory makes it possible to better elucidate the population-environment relationship, that is, analysis of the dependency and complementarity relationships between these entities, with the aim of ensuring better management of resources and optimizing perfect control of space. In the case of this study, therefore, it will be a matter of trying to grasp the impact of farmers on the transformation of the Sudano-Sahelian space through their choices of production systems, access to land and soil management practices.

7.4. Methodological approach adopted

The information and data were collected through three approaches. The first involved a diagnostic survey of soybean producers and key resource persons in the local institutional system; the second focused on changes in vegetation cover; and the third on the state of soil degradation. The study targeted 12 localities (Touboro, Madingring, Ngouni, Mbai-Boom, Mbidere, Mbissiri, Pandjama, Mbailara, Sorombéo, Mayo-djarendi, Gor, Djemadjou). These form the front of cotton expansion over the past 20 years, where most of the production dynamics for soybeans can be observed. Obtaining statistics on production required arbitration between the source of the Ministry of Agriculture through its local branches on the one hand and that of the FAO (FAOSTAT) on the other. Despite the significant differences between these two sources, the data from the branches were preferred because of their proximity to the field, notwithstanding doubts about their reliability. This choice is supported by the authors' participation in monitoring changes in the food subsector in the cotton area since 2006[11]. This year precedes the start of experimental trials by SODECOTON for crop diversification.

7.4.1. *Diagnosis with soybean producers and resource persons*

A questionnaire survey was conducted between 2017 and 2019 with 500 farmers who were members of soybean producer groups selected using the "snowball" method. The surveyed farmers had to have at least 1 ha of soybeans in one or more plots. Questionnaires were administered directly to farmers on their motivations for growing soybeans, modes of access to land, use of phytosanitary products and

11 Monitoring the evolution of the food sub-sector in the Cameroonian cotton zone was the subject of a regional research project involving Cameroon, Chad and the Central African Republic from 2004 to 2009, as part of Projet Appui, a regional search for sustainable development of Central African savannahs (see: https://afrique-ouest.cirad.fr/content/download/7141/70587/version/1/file/Plaquette_ARDESAC.pdf [Accessed February 14, 2021]); (Ndjouenkeu et al. 2010).

institutional and non-institutional technical support. In addition, interviews were conducted with officials of the Soyabeans Processing Industry of Cameroon (SOPROICAM), the agribusiness company that currently governs the soybean sector in the Sudano-Sahelian zone, and staff of the Ministry of Agriculture.

7.4.2. *Analysis of the evolution of vegetation cover*

The analysis based on Landsat images used a diachronic approach over three time steps: 2008, which marks the year of soybean extension in the Cameroonian cotton basin; 2014, which symbolizes the new era of soybean cultivation following the disengagement of SODECOTON from all forms of support for the development of the crop and the involvement of cooperative societies for the purchase of production; and 2018, which represents the year of the soil samples taken. The cartographic treatments from appropriate techniques have allowed the realization of land use maps.

7.4.3. *Soil sampling*

Soil samples were collected from the soybean fields at depths ranging from 0–30 cm to 0–50 cm. Samples were taken according to the toposequence of the study area (upper slope, slope or mid-slope and lower slope). Each sampling operation resulted in 10 soil samples from 10 previously dug holes, which were then mixed using the quartzing and soil sampling technique. One kilogram of soil was collected 30 cm from the ground, representing one composite sample per site out of approximately 3,500–4,000 tons of available topsoil per hectare. For each sample taken, optimization according to quality and cost objectives was carried out. For this purpose, important elements such as the size of the plot where the sample is taken, the type of decontamination and the desired degree of precision were taken into consideration.

Sampling areas were selected based on the criteria of total area of the producer's plot, age of the field, and the particularity of the plot. Three sites were chosen: foothills, plains and mountain tops. A working area 20 m long and 10 m wide was defined beforehand for each site. The equipment used consisted of the following:

– biodegradable plastic bags with sample codes to avoid confusion;

– a cooler to ensure homogeneity and secure transport of samples;

– protective gloves to avoid contamination of samples;

– one meter for the delimitation of the surfaces;

– a machete;

– an auger for soil perforation.

Each delicately packaged sample was identified by a permanent reference on a label with an indelible marker and attached to the plastic bag (Figure 7.1).

A

B

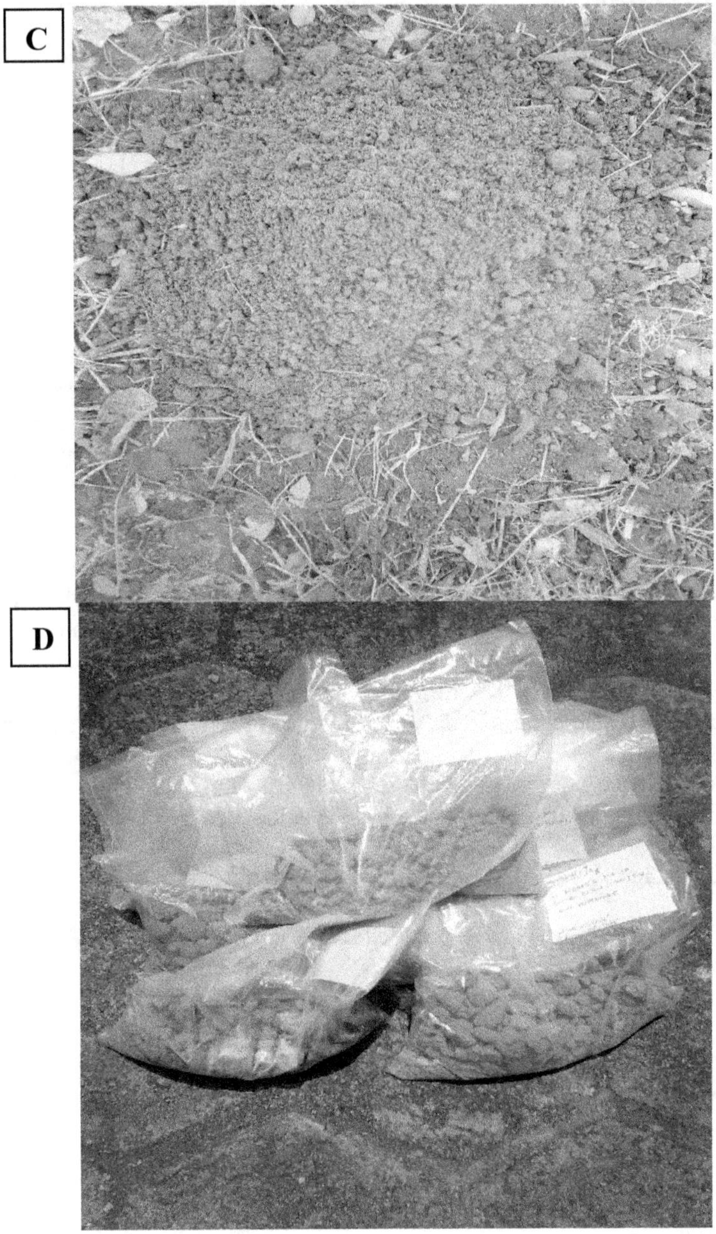

Figure 7.1. *Collection and conditioning of soil samples in Touboro and Madingring (X: N 08°3552°, Y: E 015°0116°, Z: 519 m). Images: Nitcheu Wakponou (2018). For a color version of this figure, see www.iste.co.uk/casadella/agri-innovations.zip*

The analyses were carried out with the assistance of the analytical laboratory of the Faculty of Agronomy and Agricultural Sciences (FASA) of the University of Dschang in Cameroon. Two types of analyses were performed:

– mechanical analysis or granulometry consisting of the rate of sand, silt and clay contained in the soil and the determination of the textural class;

– organic matter (OM) or physicochemical analysis (soil moisture content, phosphoric acid analysis, exchangeable bases or cation exchange capacity analysis, exchangeable acidity and hydrogen potential analysis)

Evaluation of the hydrogen potential (pH)

Soil pH was measured using a CG822 pH meter with a combined pH electrode. The actual acidity (pHH_2O) was measured in a 1:2.5 ratio soil–water suspension (10 g soil in 25 mL water) at least 16 h after preparation. Potential acidity was measured in a molar Sol-KCl suspension of the same ratio 10 minutes after preparation.

Evaluation of the organic matter

The determination of OM was performed by the Walkley and Black method, as described by Pauwels et al. (1992). This method is based on the oxidation of organic carbon by potassium dichromate ($K_2Cr_2O_7$) in a strongly acidic medium (H_2SO_4). The titration of the excess K_2Cr2O_7 by ferrous sulfate ($FeSO_4 \cdot 7H_2O$) allows the quantity of dichromate neutralized by the organic carbon (OC) to be calculated.

The equivalence point is indicated by the turn of diphenylamine [$(C_6 H_5)_2NH$] from purple to green. The percentage of OC is calculated by the following formula:

V0 = volume of $FeSO_4 \cdot 7H_2O$ added to control

V = volume of $FeSO_4 \cdot 7H_2O$ added to the sample

P = test sample (0.5 g of soil)

The OM content is derived from the following relationship:

$\% CO = 4(V0 - V) \times 100/V \cdot P$

$38 \text{ Mg N/g soil} = 14(V - V0)\, t/p$

Determination of total nitrogen

The determination of total nitrogen was performed using the Kjeldahl method. It consists of the complete mineralization of organic nitrogen by hot treatment with a mixture of concentrated sulfuric acid and salicylic acid. The mixture $CuSO_4 + S$ is used as the catalyst. Mineralization is followed by steam distillation of the nitrogen as NH_3 after alkalinization of the mineralized extract with NaOH. The distillate is captured in boric acid (H_3BO_3) and then titrated with sulfuric acid or dilute hydrochloric acid (0.01 N). The total nitrogen content is determined by the formula: V = volume of H_2SO_4 added to the sample, where V0 = volume of H_2SO_4 added to the control.

p = soil test sample in grams (2 g)

t = acid normality = 0.01 N

CEC at pH 7 (CEC7), effective CEC (ECC) and exchangeable bases

The method used allows the extraction of the exchangeable bases and the determination of the CEC from the same soil sample. It consists of three phases:

1) extraction of the bases (Ca^{2+}, Mg^{2+}, K^+, Na^+) with ammonium acetate (CH_3COONH_4) at pH 7 due to NH_4^+ ions that saturate the complex and release the basic cations (Ca^{2+}, Mg^{2+}, K^+, Na^+), which will then be measured;

2) washing of the soil with alcohol (ethanol 95%) to eliminate the saturating solution of NH_4^+ filling the porosities;

3) finally, NH_4^+ was determined by Kjeldahl distillation and titration with sulfuric acid (H_2SO_4 0.01 N) or hydrochloric acid (HCl 0.01 N). After quantitative desorption with K^+, the quantity of K^+ in solution for both the desorption studies was measured by flame photometer.

The bases extracted in the first step were determined by complexometry with EDTA (ethylene diamine tetra-acetic acid) for calcium and magnesium, while potassium and sodium were determined by flame spectrophotometry, as described by Pauwels et al. (1992). The effective CEC (ECEC) is obtained by the following formula: CEC (meq/100 g of soil) = (V – V0) t.104/Vpe. Page 39 CECE = sum of the bases + exchangeable acidity.

Assimilable phosphorus

Assimilable phosphorus was determined by the Bray II method. This method combines the extraction of phosphorus in an acidic medium (HCl 0.1 N) with the complexation by ammonium fluoride (NH4F 0.03 N) of aluminum bound to phosphorus. The determination of the extracted phosphorus was then performed by spectrophotometry with molybdenum blue using a molecular absorption spectrophotometer at a wavelength of 665 nm.

Granulometry

The weight distribution of the different textural fractions (sands, silts, clays) was determined by mechanical analysis. Sand was separated from other fractions by underwater sieving using a 50 µm sieve. Silt and clay samples were taken using a Robinson-Köhn pipette after dispersion with sodium hexametaphosphate and agitation of the suspension (silt + clay) with a rotary shaker. The textural classes were found by using the Textural Triangle of FAO once the proportions of the different textural fractions were calculated.

7.4.4. *Results*

The results present in a first articulation the impact of agrotechnological and organizational innovation (OI) on the productive dynamics of soybean in the Cameroonian cotton basin. The second articulation describes the trajectory of the construction of a value chain driven by the national agribusiness sector. The third section analyzes the environmental issues related to the production dynamics identified.

7.5. Agro-technological and organizational innovation as the basis for the development of the soybean value chain in the Cameroonian cotton front

7.5.1. *The cotton crisis, a vector for the diffusion of soybeans*

Cotton growing is at the heart of multiple issues in the Sudano-Sahelian zone of Cameroon (Figure 7.2) (Deveze 2006; Levrat 2009; Folefack 2010). The sector experienced a serious crisis from 2006 to 2011 following the constitution of large stocks of cotton fiber by China (UE-ACP 2015), leading to a drastic drop in production from 315,000 tons in 2005 to 180,000 tons in 2011 (FAOSTAT). The number of producers fell from 360,000 in 2006 to approximately 250,000 over the same period, following the drop in the purchase price of seed cotton. The consequences of this drop in the purchase price were dramatic for the producer, given the role of cotton in supporting the entire cost of agricultural inputs for food crops (cereals and legumes) (Deveze 2006; Levrat 2009, 2010; Folefack 2010). This was even more the case for SODECOTON (drastic decline in the tonnage of exported fiber and grains for the production of vegetable oil and cotton cake for animal feed).

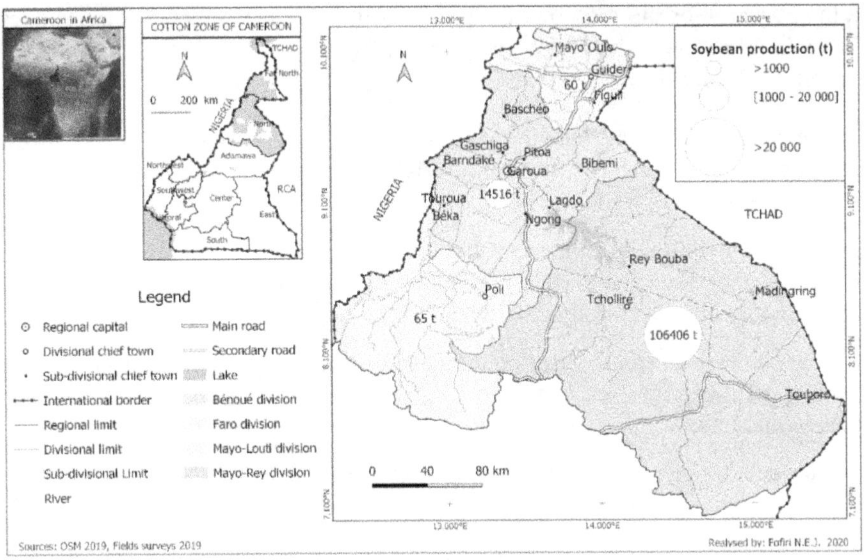

Figure 7.2. *The Cameroonian cotton front, a soybean production basin. For a color version of this figure, see www.iste.co.uk/casadella/agri-innovations.zip*

In response to this crisis[12], SODECOTON initiated a crop diversification project in 2006 through a feasibility study of soybeans and sunflowers as rotation crops with cotton (Wey and Ibrahima 2007). The results of the study indicated that soybeans[13], unlike sunflowers, could be grown as a diversification crop, mainly in the cotton front (Mayo-Rey administrative department).

SODECOTON's diversification project will be implemented starting with the 2008 agricultural season through two strategic pillars: prefinancing of agricultural inputs (fertilizers and herbicides) for soybeans exclusively for cotton producers[14]

12 This crisis is reflected in a significant variation in production between 2008 and 2011, with 185,000 tons of unginned cotton produced, 140,000 tons, 190,000 tons, and 180,000 tons, respectively, compared to 315,000 tons in 2005 (peak production that was not reached until the 2019 crop year with 32,246 tons) (data available on the FAO website: http://www.fao.org/faostat/fr/#data/QC [Accessed June 28, 2020 and February 14, 2021].

13 The so-called "local" soybean varieties (Houla1, Houla 2, TGX-849-29-4D, Carrefour Nari, Ngong and Pitoa 2) had an average yield of 1,600 kg/ha, which was more attractive than the imported IITA varieties from Ibadan and Brazil.

14 For the past 20 years, SODECOTON has adopted a policy of encouraging cotton production by providing inputs on credit for the production of food crops to cotton producers only. This policy will simply be accentuated for soybeans within the framework of the crop diversification project.

and the purchase of the crop by the agro-industry. The rapid increase in production (Figure 7.3), however, was followed by four difficulties that led to the termination of the project after 2011:

– the cessation of prefinancing of inputs and the purchase of soybean crops following the persistence of the cotton crisis in 2011 and its consequences on the financial resources of SODECOTON;

– the competition of Nigerian traders with producers to the detriment of SODECOTON, resulting in a low level of recovery of prefinancing of inputs, a situation noted in seed cotton in 2010 and 2011;

– the increasing immobilization of unsold stocks of soybean products (meal, vegetable oil), as prices are considered higher than those of cottonseed products in domestic markets;

– the technological difficulties encountered by SODECOTON in the processing of soybeans, which are more rustic than cotton beans.

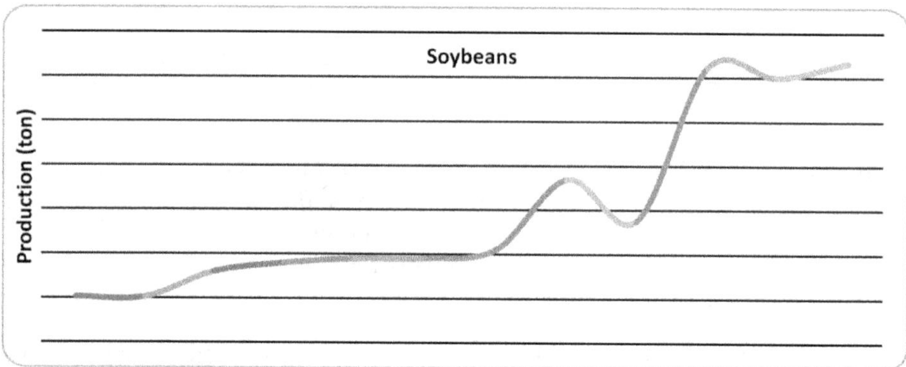

Figure 7.3. *Evolution of soybean production in the Mayo-Rey cotton front (2008–2019) (source: Regional Delegation for Agriculture and Rural Development in the North (2020)). For a color version of this figure, see www.iste.co.uk/casadella/agri-innovations.zip*

The cessation of the crop diversification support project as a result of the above-mentioned difficulties will create a new productive and commercial dynamic supported by the demand of national agribusinesses and Nigerian wholesalers. The sale of crops on domestic markets will thus become a source of direct income for producers in an environment marked by permanent cash flow tensions that cotton and other food crops are struggling to overcome. Thus, between 2011 and 2016, cotton ceased to be the main crop in rotation in favor of soybeans for 40% of the producers interviewed. This proportion was 60% after 2017, with a strong trend toward the proliferation of monoculture farms.

7.5.2. *An increase in soybean production supported more by the expansion of cultivated areas than by technical innovation*

In Cameroon's main soybean production region, the area under cultivation has increased from less than 6,000 ha in 2008 to 68,000 ha in 2019[15]. The activity brings together nearly 50,000 producers, at least 35% of whom are former cotton producers. The two crops compete for space, as well as for family and hired labor.

From a spatial perspective, soybeans are developing in the new cotton front in the Mayo-Rey department, which was opened up under the *Migration and Agricultural Support Services Program* implemented from 1974 to 1997. This program will have promoted the state-supervised settlement of more than 200,000 migrants from the Far North region on the southern shores of Lake Chad to the North (Fofiri 2013) (Figure 7.2). Both crops are therefore cultivated by these populations, who have been settled for two generations and for whom access to land remains a challenge due to persistent sociocultural barriers. The local populations have always experienced the presence of immigrants as a form of spoliation of their land (Seignobos 2006). In this context of land insecurity, the average area cultivated per producer remains modest (less than 1 ha), with nearly 50% of farms between one-fourth and three-fourths of 1 ha. This proportion is justified by the fact that soybeans are almost entirely cultivated by family farms, which also produce commodities intended primarily for self-consumption (millet/sorghum, maize, cowpeas, groundnuts). Thus, with an average yield of 1 ton/ha, the production per family farm varies between 500 and 800 kg, or five to eight bags of 100 kg. This means that production is essentially fragmented over small areas that are cultivated for an average of five years from the first year of cultivation. The rapid evolution of the cultivated areas testifies to the enthusiasm of farmers for soybean cultivation.

In some cases, soybeans are grown in rotation with corn for one to two years to capitalize on the residual effects of nitrogen. In spite of this situation, which reflects the precariousness of land for most producers, soybean monocultures of 5–15 ha are being developed by individuals and/or producer cooperatives. The reality on the

15 Source: Regional Delegation for Agriculture and Rural Development in the North (2020).

ground, however, shows a low level of mechanization of production, a consequence of the fragmentation of the farms. The tools used are mainly animal traction for plowing the fields. Less than 25% of farms use tractors for areas larger than 2–3 ha. The sowing and maintenance of crops are mainly manual operations. However, the creation of new fields through clearing and deforestation is beginning to come up against the limits of areas set aside for environmental protection (Benue and Bouba-Ndjidda national parks; zones of interest for hunting that cover nearly 50% of the northern region). The production dynamics of soybeans have not yet integrated the technical innovation that could pave the way for the development of an agro-industrial sector. The production boom is more the result of the influence of exogenous factors in the production basins than the fruit of an endogenous dynamism.

7.6. The development of the national agri-food sector: a lever for the progressive construction of a soy value chain

The paternity of the concept of value chain is recognized by Porter (1986), who developed it in his reference book *The Competitive Advantage*. The value chain describes all the activities necessary to bring a product or service from its conception through different phases of production to its distribution to final consumers, and then to its destruction after use. It allows the activity to be broken down into elementary sequences of operations and sources of potential competitive advantage to be identified. This decomposition improves the global performance of the value chain by strengthening each link and reinforcing the links between the links.

This decomposition has made it possible to analyze the organizational and functional dynamics of the soybean market, driven by growing demand from the national agri-food sector. This sector has indeed experienced a significant evolution over the last 15 years. It includes companies whose activities are related to the extraction of vegetable oil, the development of dairy products, infant flours, oil cakes and animal feed. The diagnosis of this sector identified six companies under Cameroonian law for which soybeans are one of the main raw materials in the production process (Table 7.1). The development of poultry farming has also accelerated investment in the processing of soybeans into oilcake and animal feed. However, several processing units operate informally while contributing to the increased demand of the agri-food sector.

No.	Name	Products developed	Geographic location
1	SOPROICAM	– Vegetable oil – Cakes	Douala
2	SODECOTON	– Vegetable oil – Cakes	Garoua
3	CAMLAIT	– Nouriss – Breaksoy – River	Douala
4	NT Foods S.A.R.L (Tanty, brand name)	– Soy flour – Queen Custard Powder infant soy porridge – Tanty Custard Powder	Yaoundé
5	Nkam feed mill company	– Provende (poultry feed)	Yaoundé
6	Cameroon feed mill company	– Provende (poultry feed)	Bafoussam

Table 7.1. *Agro-industrial enterprises of soybean processing in Cameroon (source: authors' compilation)*

No.	Name	Nature of the project/products developed	Geographic location
1	Entreprise Rend Foods	– Semisolid soy-based food alternatives for weaning infants (yogurts and other dairy products) under the trade name Rivoli Soy S.A. (Absorbency of raw material not specified)	Douala
2	Agrivar Cameroun S.A.	– Vegetable oil – Cakes (Crushing capacity of 150,000 tons per year, expandable to 300,000 tons if raw material)	Douala
3	PAFIC Sarl	– Vegetable oil – Cakes (Crushing capacity not specified)	Yaoundé

Table 7.2. *Agro-industrial soybean processing projects in Cameroon supported by the Cameroonian government since 2019 (source: authors' compilation)*

The vitality of this sector can also be appreciated through the intentions of agribusiness projects, some of which have been supported by the Cameroonian government since 2020 through various institutional supports and commitments materialized by the signing of partnership agreements (tax and customs exemptions over 5–10 years, as provided for by the 2013 law revised in 2017 on incentives for private investment in Cameroon) (Table 7.2).

Despite the lack of complete data to accurately assess the evolution of the national agri-food sector's needs for soybeans, two significant variables may underlie the hypothesis of technological innovation in processing: the proliferation of large-scale enterprises on the one hand and the diversification of the supply of processed products (vegetable oils, oil cakes, dairy beverages, growth foods) on the other. However, a precise understanding of this technological dynamism requires further study.

The technological dynamism of the agribusiness sector thus appears to be an important lever for building the soybean value chain, whose production follows two marketing trajectories: contractual purchase and sale on domestic markets. The first path is supported by the Soyabeans Processing Industry of Cameroon (SOPROICAM), which has set up a network of 20,000 farmers in over 2,000 producer organizations. The contractual relationship is guaranteed by the prefinancing of inputs (seeds and phytosanitary products) at the beginning of the agricultural season. In return, producers provide their crops at a price determined at the beginning of the season. The kilogram set for the 2018–2019 crop year was 170 FCFA (or 0.259 Euros), compared to 120–150 FCFA (0.182–0.228 Euros) on the domestic market. For this crop year, SOPROICAM collected 6,000 tons of soybeans (i.e. 60,000 bags of 100 kg) for a value of 1,020,000,000 FCFA (1,554,980 Euros). The unfulfilled purchase forecasts for this same season were 20,000 tons, which would have generated a money supply of 3.4 billion FCFA (5,183,270 Euros) in circulation in the region for the production segment alone.

The second path is organized around a network of wholesalers who buy from producers on rural markets. These wholesalers act as relays for national or cross-border agribusinesses. For example, the Cameroonian dairy products company (Camlait), based in the coastal city of Douala, introduced 581,203 kg of soybeans (5,812 bags of 100 kg) into its yogurt production process in 2019. This quantity represents between 69,744,360 FCFA and 87,180,450 FCFA[16]. In addition to food uses (vegetable oil, beverages, inputs for dairy products), soybeans are the main raw material used for the manufacture of feed for poultry. However, the national demand for soybean meal remains unassessed, as does the demand from Nigerian agribusinesses

16 106,325 and 132,906 Euros.

that maintain a large supply circuit on Cameroonian markets through a network of wholesalers.

The two market trajectories thus highlight evidence of the structuring of an integrated value chain that articulates four links: the supply of basic goods and services, production, distribution and processing.

The first link is built around three uncoordinated actions that nonetheless provide effective support for the provision of basic agricultural services (seeds, plant protection products and financing). SOPROICAM supplies mineral fertilizers with a ternary combination of N (nitrogen)–P (phosphate)–K (potassium) in formulations of 12–24–12 and 20–10–10, as well as herbicides (glyphosate, roundup) to producers. This company also acts as a guarantor with microfinance institutions for granting loans to producers.

However, a fragile segment of this first link remains the supply of seeds, which are essentially taken from the harvest. Since the introduction and popularization of ecologically adapted varieties by SODECOTON from 2007 to 2010[17], seed production has not been the subject of a specific research or experimentation program with convincing results. As a result, the seeds in circulation come mainly from the repeated reuse of seeds taken from harvests. In response to this weakness, SOPROICAM has initiated an improved seed production project that is gradually being structured.

The second link in the value chain includes two categories of producers. The first brings together nearly 20,000 producers in various forms of groupings (joint initiative groups, associations, cooperatives). The second is made up of independent producers. The first initiatives to organize soybean producers emerged in 2011 in the production basin. They quickly evolved into the creation of increasingly structured groups thanks to the support of SOPROICAM. The file of cooperative societies in the north region lists 10 legalized soybean producer cooperatives since 2017, six of which are in the department of Mayo-Rey alone, accounting for 80% of production in the north administrative region (Table 7.3).

Women represent 20% of the members of the groups. This low proportion can be explained by the density of labor-intensive tasks on the one hand and the immediate financial profitability provided by the sale of soybeans, which is of great interest to men, on the other hand.

---

17 So-called "local" varieties: Houla1, Houla 2, TGX-849-29-4D, ESA and SJ 235; varieties from the IITA collection in Ibadan, Nigeria: TGX 1448 2E, TGX 1485 1D and TGX 1910 14F.

Borough	Number of cooperatives	Number of members
Rey-Bouba	1	700
Tcholliré	1	1,000
Madingring	1	650
Touboro	3	2,000

Table 7.3. *Spatial distribution of cooperative societies of soybean producers in Mayo-Rey (2019) (source: Surveys of producer cooperatives and the cooperative society database of the Délégation régionale de l'agriculture et du développement rural du Nord (April 2020))*

The third link represents the heart of the value chain. It is made up of SOPROICAM and a network of Cameroonian and Nigerian wholesalers, two categories of actors that ensure the governance of the soybean sector. While it is true that the first actor has invested in controlling the sector from the field to the factory, the wholesalers are the secular arms of the processing companies, which express quantitative rather than qualitative needs (varieties, color). The fourth link is the processing companies, whose role in the construction of the value chain has already been explained at the beginning of this section.

The market relations linking the four links above confirm the existence of OI that participates in the structuring of a value chain on soybeans, OI being part of non-technological process innovations (Edquist et al. 2001; Abernathy et al. 2018). The growing investment of distribution and more processing actors in supporting the productive dynamic testifies to the challenges that currently arise from the need to sustain the national supply to meet the growing demand of the agri-food sector. However, this production dynamic needs to be supported by technical changes that are still very deficient in the current context (mechanization of plot preparation, sowing and harvesting, irrigation systems, etc.). Thus, if from an economic point of view, the progressive structuring of the soybean value chain reassures us of an endogenous dynamic capable of connecting to national, subregional, continental, and even global markets, we could nevertheless question the environmental consequences of the observed productive dynamic in light of the precariousness of the natural capital of the Sudano-Sahelian area, which constitutes the main production basin.

7.7. Environmental issues of soybean production dynamics in the Sudano-Sahelian region

The rapid increase in production and the corresponding increase in surface areas has inexorably led to the sudden and unplanned subjection of the environment, impacting in particular the vegetation and the soil.

7.7.1. *A significant regression of the vegetation cover in an ecologically fragile area*

Soybean cultivation is putting increasing pressure on vegetation cover. Analysis of vegetation dynamics in the production basins over the last 20 years has revealed a significant regression of forest, grassland and savanna areas, converted to agricultural spaces. The state of land use in the two largest production basins in 2008, 2014 and 2018 confirms trends of strong degradation of natural vegetation substituted by fields (Figures 7.3–7.5). The year 2008 marked the introduction of cultivation in the context of predominantly cotton fields and other food crops (maize, groundnuts, cowpeas, voandzou). The year 2014 corresponded to the withdrawal of SODECOTON in favor of SOPROICAM, Cameroonian and Nigerian traders. The year 2018 marked the beginning of observations within the framework of the study.

Thus, in 2008, the landscape was marked by the presence of cotton fields materialized by white shades, polarized around the migrant settlement areas. Cartographic representation was reflected in the low rate of bare soil and lush vegetation dominated by gallery forests, grassy savannahs and open forests (Figure 7.4).

From 2014 onward, when the production boom was observed, the dynamics of the environment have been marked by a gradual spread of bare soil and, conversely, a decrease in vegetation cover resulting from the creation of new fields by clearing. The trend toward bare soil is most noticeable around the villages that have been created on the Mayo-Rey cotton front over the past 20 years as part of the North–East Benue Project (Djemadjou, Djeng, Mbissiri, Pandjama, Vogzom, Mbidere) (Figure 7.5). These villages are populated by migrants with an old agricultural tradition from the Far North, in contrast to the local populations who are more oriented toward extensive cattle breeding. The spread of land clearing can be explained by two factors revealed by the producers. The first is related to the decline in soil fertility after four to five years of production. The second is underpinned by the increase in the number of soybean producers. This increase is sustained by the migration flows that continue in the "Far North–North" direction for crop land. Independent of these two categories of producers, new fields are opened up through the practice of slash-and-burn, which is highly damaging to vegetation, soils and biodiversity. The contribution of buildings to soil denudation, although significant, remains limited to the central areas of villages.

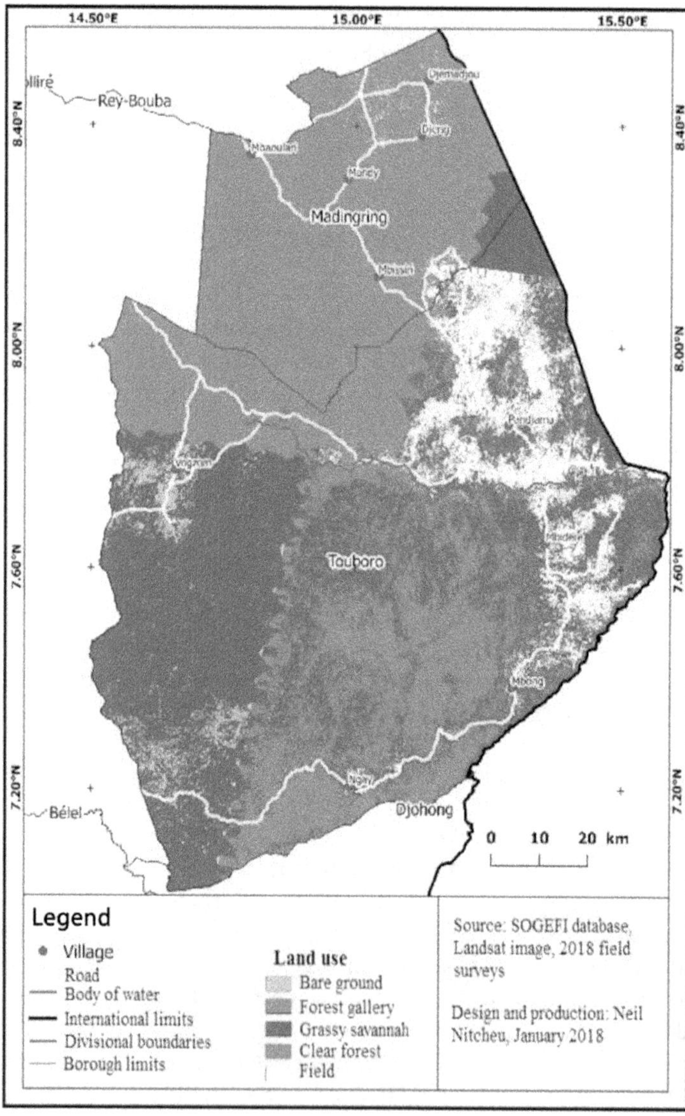

Figure 7.4. *Land use in Touboro and Madingring in 2008. For a color version of this figure, see www.iste.co.uk/casadella/agri-innovations.zip*

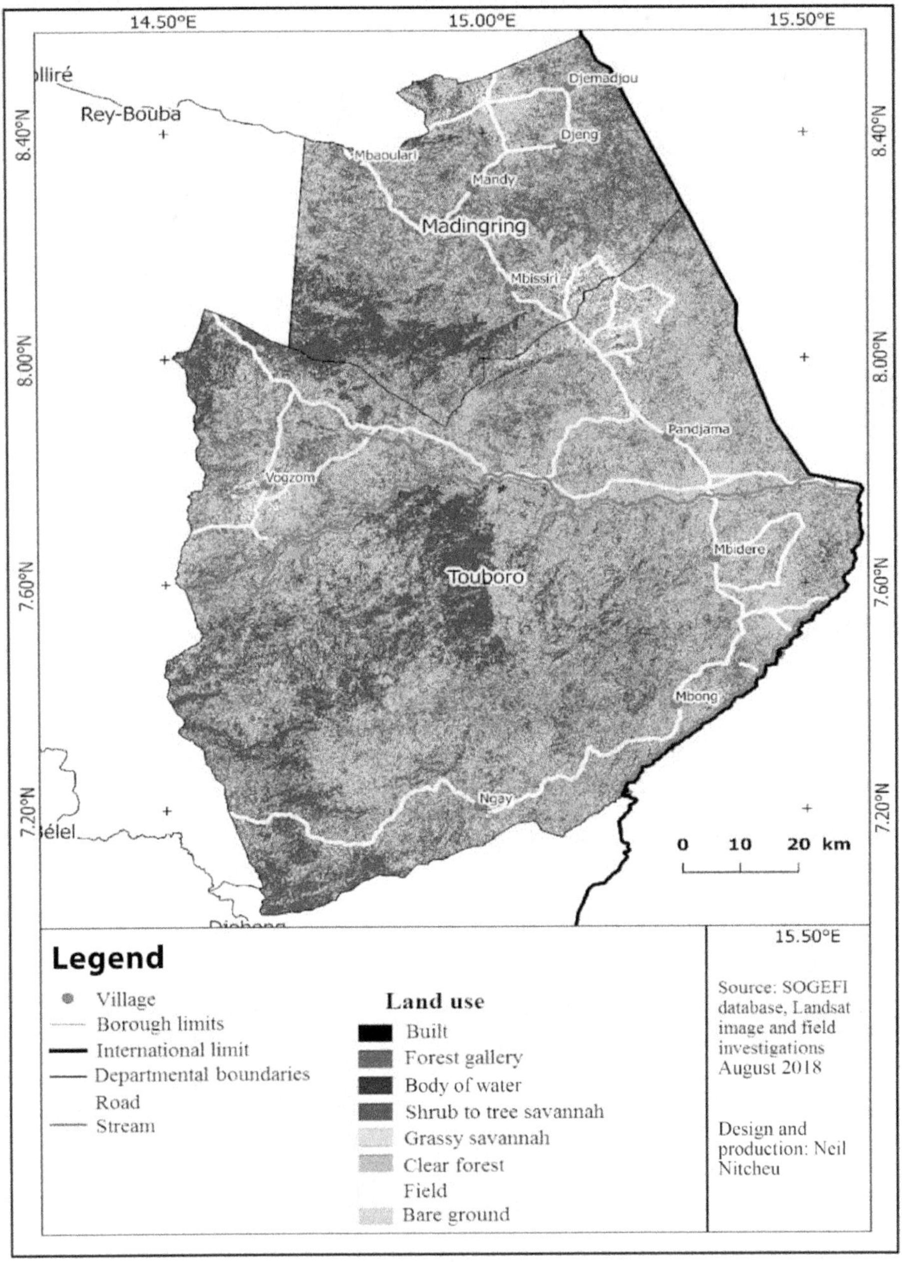

Figure 7.5. *Land use in Touboro and Madingring in 2014. For a color version of this figure, see www.iste.co.uk/casadella/agri-innovations.zip*

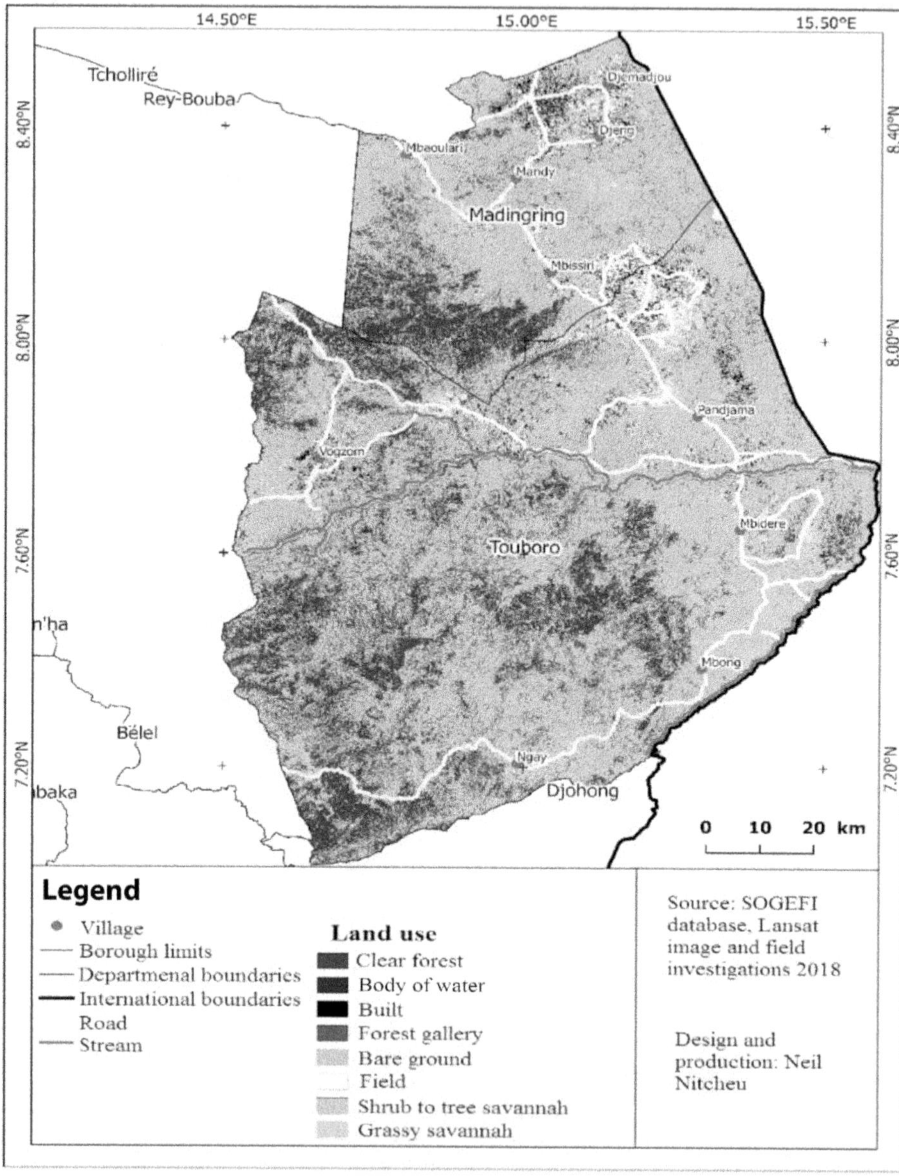

Figure 7.6. *Land use in Touboro and Madingring in 2018. For a color version of this figure, see www.iste.co.uk/casadella/agri-innovations.zip*

In 2018, however, denuded land was reclaimed by natural vegetation, albeit tentatively (Figure 7.6). Exchanges with farmers have made it possible to note that this recovery is due to the sensitization of populations on the rational use of space and the dangers of poor agricultural practices on vegetation initiated by certain local associative movements, supported by sectoral administrations (agricultural and forestry station chiefs), communes, civil society and SOPROICAM.

Overall, the analysis of the evolution of bare soil areas over the last two periods (2014–2018) reveals variable dynamics according to the areas. This evolution was particularly strong for the built-up area, gallery forests, shrub and tree savannahs and fields, which have seen a doubling of surface areas in the space of 10 years (Table 7.4). This doubling can be explained by the need for systematic weeding for soybean cultivation, particularly between the 15th and 35th days after sowing. From an agronomic point of view, the presence of weeds around the plant from the time of sowing constitutes a competition object for light, nutrients and water. Moreover, grasses constitute a spawning ground, refuge par excellence for insects that attack the plant and transmit diseases.

Domains	2014	2018
Built-up area	3,066.3	8,066
Gallery forest	31,0445	66,125.9
Lake	1,975.05	1,596.15
Clear forest	340,182	309,225
Shrubby to wooded savannah	346,529	599,549
Grassy savannah	66,384.1	69,014.2
Field	23,350.2	46,700.4
Bare soil	221,214	212,869
Total	1,313,145.65	1,313,145.65

Table 7.4. *Calculation of land cover areas for the years 2014 and 2018 (in km²)*

In addition, the main production basins (Touboro and Madingring) are home to a network of zones of hunting interest (ZIC) (Z12, Z17, Z20, Z21, Z22), as well as national parks (Boubba Djidda National Park (PNBD), Benue National Park (PNB)). The creation of these protected areas by Cameroon over the past 40 years is a materialization of the country's desire to preserve this historically precarious environment. The degradation of the vegetation cover observed is therefore a vector of annihilation of this institutional will.

7.7.2. *Soil degradation linked more to cultivation practices than to soybean speculation*

Soil degradation is a major problem in the development of food crop agriculture in the Sudano-Sahelian region. The analysis of soil samples taken from plots where soybeans are grown has made it possible to make an initial assessment of the impact of soybeans on the soil. Three significant results are highlighted in the two production basins studied: soil acidity, OM concentration and nitrogen (N) supply.

For the first result, the analysis of hydrogen potential levels in the soils revealed high acidity in Touboro, with levels reaching −1.4, and relative alkalinity in Madingring. Data from the producer surveys revealed widespread use of herbicides on farms in Touboro and Madingring. This trend to spray soybean fields is justified by the duration of action of glyphosate (the active ingredient in Roundup) on weeds, especially in new plots. Thus, 56% of respondents have systematized herbicide spraying, 22% alternate spraying and manual weeding, 18% have adopted exclusive manual weeding, and 4% practice the system under plant cover (UPC), especially on newly cleared plots. Similarly, 40% of producers use bush fires as an alternative.

However, it must be emphasized that the choice of one or the other practice remains strongly dependent on the size of the farm. In general, beyond 1 ha, manual weeding requires much labor to be paid. However, the use of herbicides remains an important factor in soil acidification. The second result showed a very low concentration of OM, characteristic of poor soils, with proportions ranging from 0.75% to 0.42%. This low concentration is justified by the precariousness of the vegetation cover in this Sudano-Sahelian area. The exacerbation of land clearing is thus a factor that amplifies the poverty of soils in OM. The third result showed a high concentration of nitrogen, with proportions ranging from 0.439 to 0.574 at Touboro and Madingring. This corroborates the recognized nitrogen production function of legumes (Beirnaert 1941; Kananji et al. 2013; Ndusha et al. 2017).

In short, the results obtained from the analysis of soybean cultivation practices on vegetation and soil reveal the scope of the environmental challenges of this crop in the Sudano-Sahelian region. In addition to socioeconomic issues, the trend toward increased production poses a risk of perverse effects on the dynamism of Sudano-Sahelian farmers by exacerbating environmental degradation.

7.8. Conclusion

In this study, we proposed to analyze the sustainability issues of the productive and commercial dynamics of soybeans, a crop with a high potential for improving farmers' incomes, but that also poses environmental degradation issues in the

Sudano-Sahelian region of Cameroon. The production dynamics described follow a spatial diffusion trajectory structured in three stages: the initial stage of appearance and growth of soybean cultivation starting in 2008, the intermediate stage of accelerated growth observed since the 2015–2016 agricultural year, and the final stage of saturation and decline. The temporality of the first two stages testifies to the rapid penetration of the environment by technical and OI, which combined varietal selection, technical itineraries, choice of soils, and marketing strategy. This speed also reflects the capacity of Sudanese-Sahelian farmers to appropriate innovations when they understand their interest and contribution to their socioeconomic development, and to optimize their use.

However, from an agronomic point of view, the low productivity (on average 1 ton/ha) reflects the weight of the constraints that limit the optimization of the productive potential of nearly 2 tons/ha, revealed by the experimental trials in 2006, mainly on local varieties. The study thus identifies a series of weaknesses and constraints that affect both small and medium-sized farms and large farms. In spite of these weaknesses and constraints, the evolution of production based on a trajectory of spatial spread offers the opportunity to build a value chain that articulates strategic links.

While this development can be expected to improve farmers' incomes in light of the proposed evaluation of the economic profitability of soybeans, the cultivation practices adopted by farmers have environmental consequences, in this case on the plant cover, the soil and the loss of biodiversity. The results obtained corroborate the postulate that the desire of producers to meet the growing demand for soybeans in a context of low production intensification is based on the trajectory of opening up new cultivation fronts increasingly within the limits of protected areas, which exacerbates the clearing of vegetation cover and induces the massive use of chemical substances to ensure the protection of crops on plots. This context of increasing environmental fragility is in itself a precarious situation for the sustainability of the soybean value chain in the Cameroonian cotton front.

To this end, the study proposes to integrate the landscape restoration approach, which takes into account the logic of productivity, while respecting the sustainability of the environment. This guides public action toward choices based on the adoption of the best practices likely to preserve the sustainability of the soybean value chain.

The observation of environmental fragility also requires further reflection and the linking of this threat to soils with the issue of forest landscape degradation, for which Cameroon has committed to restoring approximately 12 million hectares of degraded landscapes and land. From this perspective, the efforts envisaged should integrate the limitations of the current factors causing the degradation of land resources. This requires the characteristics of landscape restoration to be taken into account, namely:

– adoption of the participatory process based on social, technical/technological, economic and environmental changes;

– restoration of the ecological integrity of ecosystems;

– improvement of human well-being.

The limitations of this study relate to the difficulty of accurately characterizing the technological changes that underlie the diversification of product offerings by agri-food firms. This difficulty is due to the sensitivity of the information on technological innovations requested from firms in a highly competitive national industrial environment. The second limitation relates to the difficulty in quantifying cross-border soybean export flows due to the illicit nature of agricultural trade between Cameroon and Nigeria (Fofiri et al. 2018).

7.9. References

Abernaty, W.J. and Utterback J.M. (1978). Patterns of industrial innovation. *Technology Review*, 80, 40–47.

Aragrande, M. (1997). Les approches disciplinaires de l'analyse des SADA. Speech presented at the FAO-ISRA sub-regional seminar "Approvisionnement et distribution alimentaires des villes de l'Afrique francophone", Dakar, 14-17 April. "Aliments dans les villes" collection.

Arvor, D. (2009). Etude par télédétection de la dynamique du soja et de l'impact des précipitations sur les productions au Mato Grosso (Brésil). Sciences de l'Homme et Société. Université Rennes 2.

Arvor, D., Dubreuil, V., Mendez del Villar, P., Magri Ferreira, C., Simões Penello Meirelles, M. (2020). Développement, crises et adaptation des territoires du soja au Mato Grosso : l'exemple de Sorriso. *Confins* [Online]. Available at: http://journals.openedition.org/confins/5934 [Accessed 10 July 2020].

Baboy, L., Kidinda, L., Kilumba, M., Langanu, S., Mazinga, M., Tshipama, D., Kimuni, L. (2015). Influence du semis tardif sur la croissance et le rendement du soja (Glycine max) cultivé sous différents écartements à Lubumbashi. RD Congo. *International Journal of Innovation and Applied Studies*, 1, 104–109.

Beirnaert, G. (1941). La technique culturale sous l'Equateur. Report, INEAC.

Casadella, V. and Tahi, S. (2021). Constat et analyse des politiques industrielles et scientifiques dans le contexte de l'intégration régionale du Sénégal. *Revue science, technologie, développement*, 21–1(1), 14.

Devèze, J.C. (2006). Le coton, moteur du développement et facteur de stabilité du Cameroun du Nord ? *Afrique contemporaine*, 217, 107–120.

Dronne, Y., Forslund, A., Gohin, A., Guyomard, H., Levert, F. (2007). Impacts du développement des biocarburants aux États-Unis et dans l'UE sur les marchés internationaux de produits de grandes cultures. *Oilseeds and Fats, Crops and Lipids (OCL),* 14(6), 347–353.

Dubouloz, S. and Bocquet, R. (2013). Innovation organisationnelle. S'ouvrir pour innover plus ? *Revue française de gestion*, 235, 129–147.

Dumolard, P. (1999). Accessibilité et diffusion spatiale. *Espace géographique*, 28(3), 205–214.

Edquist, C., Hommen, L., McKelveg, M. (2001). *Innovation and Employment: Process versus Product Innovation*. Edward Elgar, Cheltenham.

Fofiri Nzossie, E.J. (2013). *Les déterminants de l'offre alimentaire vivrière dans les villes du Nord-Cameroun*. PhD Thesis, Université de Ngaoundéré.

Fofiri Nzossié, E.J. and Bring, C. (2020). Soybean (*Glycine max L.* Merr.) production in the Cameroonian cotton basin between the dynamics of structuring an agricultural value chain and sustainability issues. In *Soybean for Human Consumption and Animal Feed*, Sudarić, A. (ed.). IntechOpen.

Fofiri Nzossie, E.J., Temple, L., Ndjouenkeu, R. (2015). La contribution de la recherche universitaire à la formation d'un système sectoriel d'innovation agro-alimentaire au Cameroun. *Cahiers d'économie de l'innovation/Journal of Innovation Economics*, 47, 55–77.

Fofiri Nzossie, E.J., Temple, L., Ndame, J.-P. (2018). L'impact des échanges agricoles transfrontaliers illicites sur la sécurité alimentaire urbaine au Nord-Cameroun. In *La gouvernance urbaine en Afrique. Quelles priorités ?*, Simeu Kamdem, M. (ed.). Editions Kathala, Paris.

Folefack, D.P. (2010). Coordination des acteurs dans un contexte de crise : le cas de la filière coton au Cameroun depuis 1990. Economies and finances. Université Rennes 2, Université Européenne de Bretagne.

Folefack, D.P., Ntsou Bakwowi, J., Cokou Kpade, P.J. (2014). La crise de la filière cotonnière et sécurité alimentaire au Nord Cameroun. *Journal of Applied Biosciences*, 75, 6221–6231.

Gendreau, F., Gubry, P., Véron, J. (1996). *Populations et environnement dans les pays du Sud*. Éditions Karthala, Ceped, Paris.

Gould, P., Dibiase, D., Kabel, J. (2014). Le SIDA la carte animée comme rhétorique cartographique animée. *Mappemonde*, 1, 21–26.

Griffon, M. (2003). Quand l'agriculture africaine va-t-elle commencer à répondre aux enjeux du futur ? *Cahiers agricultures,* 12, 141–143.

Guéneau, S. (2018). Durabilité des chaînes globales de valeur du soja et de la viande de bœuf en Amazonie : conséquences d'une gouvernementalité néolibérale. *Brésil(s)* [Online]. Available at: http://journals.openedition.org/bresils/2468 [Accessed 23 June 2020].

Gurtler, J.-L. (2008). Biocarburants : développement et perspectives. *OCL*, 15(4), 235–239.

Hàgerstrand, T. (1952). The propagation of innovation waves. In *Lund Studies in Geography, Ser. B, no. 4*. Royal University of Lund, London.

Hàgerstrand, T. (1957). Migration and area. In *Lund Studies in Geography, Ser. B, no. 13*. Geographical Information, London, 223–224.

Hart Energy (2013). Global biofuels outlook to 2025. Global Biofuels Center [Online]. Available at: https://ubrabio.com.br/sites/1800/1891/PDFs/InformacaodoSetor/2013Biodiesel Primer.pdf.

Kananji, G., Yohane, E., Siyeni, D., Kachulu, L., Mtambo, L., Chisama, B., Mulekano, O. (2013). *A Guide to Soybean Production in Malawi*. Department of Agricultural Research Services (DARS), Lilongwe.

Labonne, M. (1987). Sur le concept de la filière en économie agro-alimentaire. In *Terres, comptoirs et silos : des systèmes de production aux politiques alimentaires*, Kermel-Torres, D. and Roca, P.J. (eds). ORSTOM, Paris.

Le Tourneau, F.-M. (2004). Jusqu'au bout de la forêt ? Causes et mécanismes de la déforestation en Amazonie brésilienne. *Mappe Monde*, 75, 1–25.

Levrat, R. (2009). *Le coton dans la zone franc depuis 1950. Un succès remis en cause*. Editions L'Harmattan, Paris.

Levrat, R. (2010). *Culture commerciale et développement rural. L'exemple du coton au Nord-Cameroun depuis 1950*. Editions L'Harmattan, Paris.

Maudet, J.-B. (2011). La diffusion spatiale à l'épreuve des rodéos. Les hybridations tauromachiques de l'Europe à l'Amérique. *Ethnologie française*, 41, 667–675.

Mehmet, O. (2008). Nitrogen rate and plant population effects on yield and yield components in soybean. *African Journal of Biotechnology*, 7(24), 4464–4470.

Ndusha, B., Karanja, N., Woomer, P., Walangululu, J., Mushagalusa, G., Sanginga, J.M. (2017). Effectiveness of rhizobia strains isolated from South Kivu soils (Eastern DRCongo) on nodulation and growth of soybeans (Glycine max). *African Journal of Soil Science*, 5(3), 367–377.

Nitcheu Wakponou, D.N. (2018). Les enjeux environnementaux du développement de la chaîne de valeur soja dans le bassin cotonnier du Nord-Cameroun : cas des arrondissements de Touboro et de Madingring. Master's Thesis, Université de Ngaoundéré.

Nyahnone, T.J. (2017). Le développement de la culture du soja dans les fronts d'extensions cotonniers du Nord-Cameroun : le cas du Mayo-Rey. Master's Thesis, Université de Ngaoundéré.

Ouharon, A. (2006). Population et environnement : des liaisons incertaines. *Natures sciences sociétés*, 14, 168–173.

Saint-Julien, T. (1985). La diffusion spatiale des innovations. Report, GIP RECLUS, Montpellier.

Saint-Julien, T. (1992). Diffusion spatiale. In *Encyclopédie de géographie*, Bailly, A., Ferras, R., Pumain, D. (eds). Economica, Paris.

Sandron, F. (1992). Population et environnement : le paradigme de la complexité. Travaux et Documents, n°2, Programme 4D, Institut Catholique de Madagascar et Institut de Recherche pour le Développement.

Schneider, A. and Huyghe, C. (eds) (2015). *Les légumineuses pour des systèmes agricoles et alimentaires durables*. Editions Quae, Versailles.

Scott, G. and Griffon, D. (1998). *Méthodes pour analyser la commercialisation agricole dans les pays en développement*. CIRAD, Montpellier/Karthala, Lima, Paris.

Seignobos, C. (2006). Colloque international SFER "Les frontières de la question foncière". Cirad, Montpellier.

Temple, L., Lançon, F., Montaigne, E., Soufflet, J.-F. (2009). Introduction aux concepts et méthodes d'analyse de filières agricoles et agro-industrielles. *Economies et Sociétés. Série AG Systèmes agroalimentaires*, 11, 1803–1812.

Terpend, N. (1997). *Guide pratique de l'approche filière. Le cas de l'approvisionnement et de la distribution des produits alimentaires dans les villes*. FAO, Rome.

Wey, J. and Ibrahima, S. (2006). Cultures de diversification : étude de faisabilité du soja et du tournesol dans la zone cotonnière du Nord Cameroun. Report, CIRAD.

WWF (2014). *The Growth of Soy: Impacts and Solutions*. WWF International, Gland, Switzerland.

Yankou, D. (2014). Dynamique du prix international du coton : aléas, aversion au risque et chaos. *Recherches économiques de Louvain*, 80, 53–86.

8

Impact of Good Agricultural Practices on Cashew Nut Crop Yields in Senegal

8.1. Introduction

Senegalese agriculture employs 73.8% of the rural population and has 755,532 farms (ANSD 2014). This is why sustainable improvement of agricultural productivity is at the heart of the Emerging Senegal Plan (*Plan Sénégal Émergent* (PSE))[1]. Agriculture occupies a prominent place in the economies of developing countries. To illustrate this point, Irz et al. (2001) estimate that an increase in the productivity of agricultural plots can reduce the number of poor people in Africa by 7%. Despite its importance in the fight against poverty and food insecurity, Senegalese agriculture contributes only 9.4% to the gross domestic product (GDP) (ANSD 2020a). With a coverage rate of 71% in 2018 (ANSD 2020a), Senegal is forced to resort to massive cereal imports, which cause a trade deficit of 16% (ANSD 2011). The sector is experiencing difficulties including soil degradation and impoverishment, underequipment of producers, insufficient and outdated hydroagricultural structures and a lack of production, processing/preservation and marketing infrastructure. These factors generate low agricultural productivity and poverty in rural areas and increase the exodus of rural populations to cities. The resulting densification of urban areas limits the local supply of food products and worsens the country's food insecurity. This situation creates a chronic imbalance between supply and demand for agricultural products. To reverse this negative trend, Senegal needs to transform the agricultural sector for growth that benefits rural households. It must diversify its agricultural supply and rely on sectors that offer productivity opportunities, such as the cashew sector. The cashew tree – the tree from which cashew nuts are derived – which was initially considered a reforestation

Chapter written by Blaise Waly BASSE, Souleymane MBAYE and Omar DIOP.

1 This plan is currently the reference for public policies for the period 2014–2035.

tree to combat erosion, is now a cash crop for West African countries. This is thanks to a favorable international environment, incentive prices and the development of concurrent trade relationships between West Africa and Asia (Rongead 2013; Koffi and Oura 2019). With a growing global demand for cashew, the industry is a source of income for producers (Nugawela et al. 2006; Rongeand 2013; Dieng et al. 2019). Today, out of a global production of 3,434,000 tons of raw cashew nuts, the share of African countries represents 38% (OIT 2020). In West Africa, Côte d'Ivoire is the leading producing country, with a record 800,000 tons in 2018 (OIT 2020). Senegal accounts for only 0.8% of global production (Dieng et al. 2019). Thus, it is important to identify factors that could increase production. The purpose of this research is to assess the impact of good agricultural practices (GAPs) on cashew productivity. Section 8.2 presents a review of the literature to better understand the channel through which the adoption of GAPs will affect production. Section 8.3 presents the hypothesis and the method for identifying the impact of GAPs. The results on producer characteristics and the impact of GAPs are presented in section 8.4. Finally, section 8.5 highlights the synthesis and the implications for science and technology and economic policy.

8.2. Literature review

The sustainability of agricultural growth is a major challenge for developing countries. The instability of production is explained by a combination of several factors, such as constraints in access to technology and innovation, declining soil fertility, financing difficulties and the effects of climate change (Deressa et al. 2009; Fe and Aklobessi 2018). Some authors (Awotide et al. 2011; Basse 2015; Ndiaye and Kabou 2021) consider the introduction of agricultural technologies and innovations as a tool for improving productivity. Thus, producers must be informed about the characteristics of these agricultural innovations and technologies to induce them to change their behavior. In Peru, Godtland et al. (2004) show that farmers who participate in a field school program have significantly more knowledge of integrated pest management practices. This channel of training through the field-school approach may be relevant, as there is ample evidence of a positive relationship between GAPs and agricultural productivity (Panco et al. 2013). In cashews, Balasimha (1991) showed that high saturation irradiation was necessary for an optimal photosynthetic system. This result brings the possibility of increasing cashew productivity by increasing the canopy area and increasing the photosynthetic capacity (Balasimha and Yadukumar 1992). Other authors (Barros et al. 1984; Gosh 1989; Chrisostomo 2005) discuss the importance of mineral elements in the cashew production process. However, for a better improvement of production, these authors recommend taking into account the age of the plant, the genotype used, the cultivation conditions (soil and climate) and the management of the crop.

Cassman et al. (2003) view agriculture as a resource-intensive enterprise that impacts environmental quality. Thus, for sustainable agriculture that meets human needs and protects natural resources, these authors recommend avoiding crop expansion into natural ecosystems, increasing nitrogen use efficiency and improving soil quality. However, Jelsma et al. (2019) argue that it is large farms that have the capital to implement good practices. By examining oil palm fertilizer use, planting equipment and methods, and other management practices, these authors show that these good practices do not improve productivity and sustainability of the sector in Indonesia. According to Benny et al. (2022), the main mechanism is to enable smallholders to make greater use of modern agricultural inputs and practices. In Nepal, Baral et al. (2021) explored four production technologies for ginger cultivation. Using a randomized complete block design, ginger yield following GAPs and without harvesting the mother rhizome was 17.9 tons/ha or 39.8% higher than traditional practices. In the same dynamic, Awio et al. (2022) showed that delays in weeding and fertilization created yield losses of 5.3 and 1.9 g/m^2 per day of delay, respectively.

8.3. Methodology

8.3.1. *Presentation of the study area*

The low production of cashew trees in the Kolda region is due to a combination of several factors, such as the lack of means for fertilizing the plantations, insufficient technical supervision, low-productivity varieties, recurrent bush fires, animal rambling and pests (Sarr 2002). Strategies must be considered to encourage cashew producers to adopt good practices that can improve productivity. It is within this framework that the International Relief and Development (IRD) launched the Improving the Cashew Value Chain in Senegambia project in 2014 to promote information exchange through participatory training. In Senegal, the communes of Bagadadji and Sare Bidji were selected to participate in the project.

The Champ-École model, which is a group of 25 producers, was chosen to raise awareness of GAPs. Through the field-school model, two local facilitators were made available to encourage them to respect the spacing of at least 10 m between plants and 10 m between rows, to prune the trees regularly, to put up a fence or a firewall to fight against bush fires, to apply thinning or weeding in the orchard, to combine their cashew plants with annual crops or to plow the orchard. However, the application of these GAPs is not obvious and will depend on the degree of motivation of producers to change their behavior by abandoning old practices. Since these practices are not perfectly substitutable, they must all be applied to increase productivity. Thus, all of these agricultural practices will be evaluated. In this case, any producer who has applied all of these farming practices in his orchard is considered an adopter.

This research was conducted in the Kolda region using 2018 data. It is located between 12°20 and 13°40 north latitude and 13° and 16° west longitude and covers an area of 13,721 km², or 7% of the national territory. Located in the agroecological zone of Casamance (Figure 8.1), the Kolda region is bordered to the north by Gambia, to the south by the two Guineas (Bissau and Conakry), to the east by the Tambacounda region and to the west by the Sédhiou region. It is composed of the departments of Kolda, Vélingara and Médina Yoror Foulah.

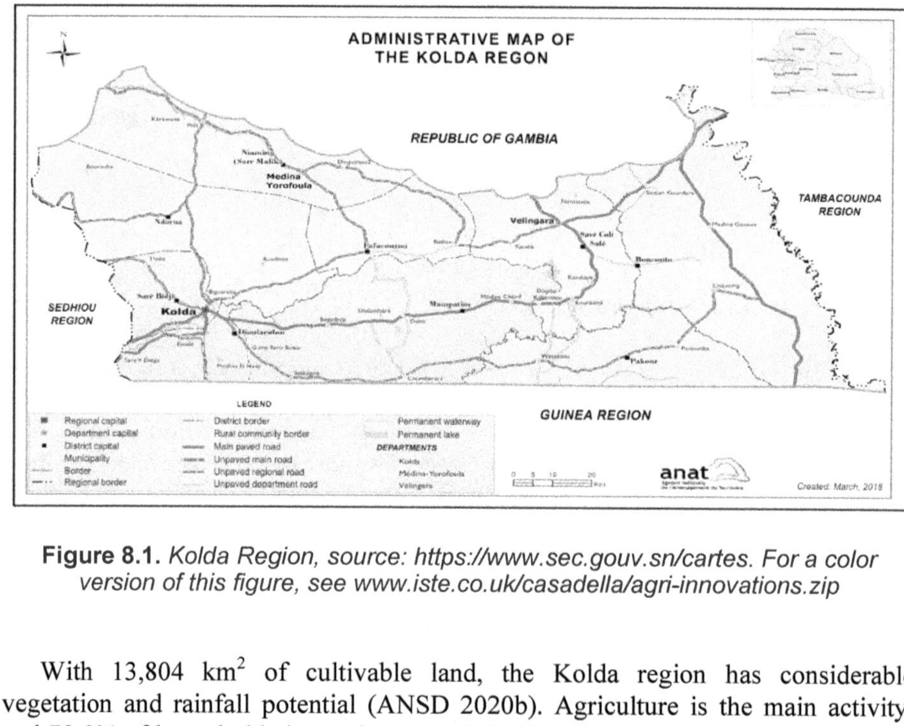

Figure 8.1. *Kolda Region, source: https://www.sec.gouv.sn/cartes. For a color version of this figure, see www.iste.co.uk/casadella/agri-innovations.zip*

With 13,804 km^2 of cultivable land, the Kolda region has considerable vegetation and rainfall potential (ANSD 2020b). Agriculture is the main activity, and 79.8% of households in rural areas and 20.2% of households in urban areas are active in the sector and derive their sources of income from it (ANSD 2014). It is composed of the rural communes of Kolda, Dabo, Bagadadji and Dioulacolon.

Pulaar is the most widely spoken language in Kolda, followed by Wolof, Manding, Sarakholé, Soninké and Diola. The department of Kolda is bordered to the north by the department of Medina Yoro Foulaah, to the east by the department of Vélingara, to the west by the region of Sédhiou and to the south by Guinea Bissau and Guinea Conakry. The climate is of the Sudanese-Guinean type with rainfall from June to October, with a maximum intensity in August and September.

The average rainfall varies between 700 mm and 1,300 mm. The lowest average monthly temperatures are recorded between December and January and vary between 25 and 30°C, and the highest are noted between March and September with variations of 30–40°C. The relief is made up of sandstone-clay forming plateaus with abundant natural vegetation (savannah or clear forest), interspersed with valleys in which rice fields and lowland pastures are found.

Bush fires are one of the causes of the degradation of forest formations. Frequent in the dry season, they ravage several hectares of forests, pastures and tons of crops every year. These bushfires have caused several damages, including the loss of plantations and crops, vegetation (burned herbaceous cover and tree vegetation), agricultural products (groundnuts, millet, rice, maize, cashew nuts), houses and burned equipment. The specific biological richness of the area is both an asset and a constraint, as some species that develop in this biotope are real crop pests.

In the orchards of the department of Kolda, the absence of fences facilitates and causes the roaming of animals and the theft of cashew nuts. One of the main constraints of the cashew nut sector is the great instability of producer prices.

In the early 1980s, 1 kg of mahogany nuts was traded for one kilo of rice at the Guinea–Bissau border. Between 1994 and 1998, prices varied, on average, between 150 F CFA (0.23 EUR) per kilo at the beginning of the season and 475 F CFA (0.72 EUR) at the end of the season. From 1999 to 2001, prices fell significantly. For the 2021 crop year, the price per kilo is set at 305 F CFA (0.46 EUR). It is clear that improving cashew nut productivity can be a determining factor in improving the living conditions of producers in the region.

8.3.2. *Sampling and data collection*

In this research, multistage stratified sampling was used:

1) Step 1: stratify the villages into two groups: the school field villages in the commune of Bagadadji, on the one hand, and the other non-school field villages in the commune of Sare Bidji, on the other hand, which we call comparison group villages.

2) Step 2: randomly select an equal number of villages from the n_{ce} villages from the group of school field and non-school field villages, respectively (so there will be

$2n_{ce}$ villages selected). The non-school field villages in Sare Bidji will be selected so that they are located relatively far from the school field village in that district.

3) Step 3: in each selected village, choose a number of cashew n_{hce} of cashew producers (so, n_{hce} is the cluster size and $n = 2 \times n_{ce} \times n_{hce}$ is the total sample size of producers in Kolda Department).

As a sampling frame, we used data from the baseline survey of the cashew value chain improvement project in Senegambia that was conducted during the months of June to October 2013 and involved 1,500 cashew producers. Then, an evaluation of the feasibility of our survey design was carried out according to the availability of our financial, material and time resources to conduct the surveys using conventional values of statistical power of 80%, significance level of 5%, with a margin of 10% and a minimum detectable effect of 10%.

The sample design and minimum sample size calculation lead to a random cluster sampling of six IRD-supported villages and six non-IRD-supported villages with a cluster size of 19 producers. These choices lead to a total sample size of 228 households (103 for each of the treated/non-treated groups), including a 10% margin forecast. With a statistical power of 80% and a significance level of 5%, this number of clusters, cluster size and total sample size led to a relative minimum detectable effect (MDES) of 27.54% on yield.

Calculating the project's performance based on the minimum detectable effect of 27.54% results in an average increase in producers' yields of 68.85 kg/hectare for adopters compared to non-adopters. This effect can be considered minimal insofar as it makes it possible to achieve less than 10% (9.18%) of the objective that the project has set.

For data collection, preliminary field visits allowed us to meet with resource persons and actors in the cashew nut sector to test the questionnaire with fifteen producers. This test validated the questionnaire in terms of both its relevance and feasibility. After the validation of the collection tools, the period from February 2 to March 2 2019 was selected to collect data related to the 2018 campaign. Indeed, after discussions with project stakeholders and local facilitators, this period was considered the most favorable for mobilizing all the producers concerned by the survey and having all the agricultural information related to the 2018 campaign. Indeed, this period coincides with the end of the winter season and slightly before the start of preparations for the 2019 cashew season. This meant that producers had time to answer questions without major interruptions. The interviews lasted an average of 45 minutes. The information collected was related to producer identification, household and orchard characteristics, income and expenses generated by cashew trees, cashew production practices, and the producer's assessment of the evolution of his orchard's yield.

8.3.3. *Data analysis method*

To determine the impact of the adoption of GAPs, we will consider Rubin's causal model (1974). This model allows us to define two potential returns for each cashew producer: Y_1 when he applies the good practices and Y_0 when he does not. More specifically, we are interested in the effect of adoption, which is a binary variable representing the status of the producer, with $T = 1$ signifying treatment and $T = 0$ non-treatment. In this case, we can rewrite the observed outcome for a given producer as a function of the two potential outcomes and the treatment status.

$$Y = TY_1 + (1 - T)Y_1 \qquad [8.1]$$

The adoption of GAP generates additional costs, particularly in terms of time and labor. Producers will make a trade-off and calculate the opportunity cost of not following GAPs. In this case, each producer anticipates the return he or she would get with or without good practices and decides whether or not to adopt them. The adoption of GAPs is an individual decision that generates a problem of selection bias and endogeneity (self-selection). In this sense (endogenous treatment), conditional or unconditional independence is an unrealistic assumption; therefore, the most plausible assumption is *selection on unobservables*, and the parameter identified is the local average treatment effect (LATE). This parameter will allow us to identify the impact of GAPs on the subpopulation of producers who were led to abandon the old practices. For this reason, to eliminate both the bias induced by observable and unobservable characteristics and to address the endogeneity problem of the participation variable, the instrumental variable method will be used (Imbens and Angrist 1994; Abadie 2003). This method assumes the existence of at least one variable Z that explains the adoption status of good practices but does not directly affect returns. For example, in the case of the introduction of GAP in the study areas, IRD used awareness raising as an instrument to trigger behavioral changes at the producer level. This instrument can induce exogenous variation (Abadie 2003), and its use ensures that its variation will allow us to identify the LATE that corresponds to the change achieved by a given policy (Heckman 2010). The use of the awareness variable as an instrument is relevant because it allows producers to know and have all useful information related to the adoption of GAPs. Indeed, Daberkow and McBride (2003) consider that awareness explains the adoption decision of agricultural technologies. From this point of view, it can induce behavioral change in cashew producers (this is the *relevance of the instrument*). This instrument, awareness, cannot directly affect cashew yields (this is *restriction by exclusion*). Producers must change their behavior and adopt GAPs so that there is a variation in their yields. We also consider that it will be difficult for a cashew producer to adopt GAPs with which he is not familiar. Thus, the awareness instrument will induce a change in behavior in the same direction (this is the *monotonicity of the instrument*). Different estimators based on the instrumental variable are available depending on functional form assumptions and assumptions about the instrument (awareness in this research) and unobserved heterogeneities (Heckman 2010). Thus, two instrumental variable-based estimators are typically used to estimate the LATE. The first is the simple non-parametric *Wald* estimator that requires only the outcome variable *Y* observed, the treatment status T and the instrument *Z*, which is randomly distributed in the producer population.

$$E(Y_1 - Y_0 | T = 1) = LATE = \frac{cov(Y,Z)}{cov(T,Z)}$$

$$E(Y_1 - Y_0 T = 1|) = LATE = \frac{\mathrm{E}(\mathrm{Y}|\mathrm{Z}{=}1) - \mathrm{E}(\mathrm{Y}|\mathrm{Z}{=}0)}{\mathrm{E}(\mathrm{T}|\mathrm{Z}{=}1) - \mathrm{E}(\mathrm{T}|\mathrm{Z}{=}0)} \qquad [8.2]$$

This parameter estimates the effect of adoption in the subpopulation of individuals (the *compliers*) who have been induced to change their behavior through an exogenous variable that satisfies the instrument's exclusion (awareness) (Imbens and Angrist 1994). Abadie (2003) showed that if the conditions of an instrument are satisfied and in the absence of covariates, we have:

$$E(Y_1 | T_1 > T_0) = \frac{E(Y.T | Z = 1) E(Y.T | Z = 0)}{E(T | Z = 1) - E(T | Z = 0)} \qquad [8.3]$$

$$E(Y_0 | T_1 > T_0) = \frac{E(Y.(1-T) | Z = 1) - E(Y.(1-T) | Z = 0)}{E((1-T) | Z = 1) - E((1-T) | Z = 0)} \qquad [8.4]$$

The second estimator based on the instrumental variable is given by Abadie (2003), which is a generalization of the Imbens *LATE* estimator (Imbens and Angrist 1994) in cases where the instrument is not totally independent of potential outcomes Y_1 and Y_0. This estimator does not require the random hypothesis but only the conditional independence hypothesis: the instrument *Z* is independent of potential outcomes. When the assumptions of the instrument are fulfilled, equations [8.3] and [8.4] become:

$$LATE = E(Y_1 - Y_0 | X, complier) = \frac{\mathrm{E}(\mathrm{Y}|\mathrm{X},\mathrm{Z}{=}1) - \mathrm{E}(\mathrm{Y}|\mathrm{X},\mathrm{Z}{=}0)}{\mathrm{E}(\mathrm{T}|\mathrm{X},\mathrm{Z}{=}1) - \mathrm{E}(\mathrm{T}|\mathrm{X},\mathrm{Z}{=}0)} \qquad [8.5]$$

Then, in the case where the covariates are continuous, the estimation of *LATE* can be performed using the non-parametric method. However, this method does not facilitate the interpretation of the results, and the accuracy of the estimators remains to be verified. Another alternative is the parametrization of the four conditional means of [8.5] as nonlinear functions of (X) to estimate $LATE = E(Y_1 - Y_0 | X, complier)$. However, how can *compliers* (potential adopters) be identified since they are not observable? Only $E(Y|X,T)$ is observed but does not allow for a causal interpretation since T is endogenous. To identify these *compliers*, Abadie (2003) used an estimation based on the weight k defined as follows:

$$k = 1 - \frac{T.(1-z)}{P(Z=0|X)} - \frac{Z(1-T)}{P(Z=1|X)} \qquad [8.6]$$

For any function $g(Y, T, X)$, Abadie (2001) has shown that:

$$E[g(Y,T,X)|T_1 > T_0] = \frac{1}{P(T_1 > T_0)} E[k.g(Y,T,X)] \qquad [8.7]$$

where $E[k.g(Y,T,X)]$ is the average of $g(Y,T,X)$ for the whole population and $P(T_1 > T_0)$ is the proportion of *completers* in the total population.

Knowing $P(T_1 > T_0)$ and k, equation [8.8] can be specified as follows:

$$E(Y_1 - Y_0 | X, T_1 > T_0) = E(Y|X, T_1 > T_0) = \alpha_0 + \alpha_1 T + \gamma TX \qquad [8.8]$$

where $T = 1$ if the producer adopts the technology and $T = 0$ otherwise, X is the vector of sociodemographic variables.

β and γ are the parameters to be estimated and $LATE = \alpha_1 + \gamma X$. To estimate equations [8.7] and [8.8], we must first estimate the rate of potential adopters $P(T_1 > T_0)$ using a standard probit model; then, equation [8.8] will be estimated by the instrumental variables method (Abadie 2003). In this research, we choose an exponential response function with and without interaction to ensure both positivity of the results and heterogeneity of the treatment effect across the subpopulation of potential GAP adopters. Following Abadie (2003), LATE is estimated with the impact command in Stata 13. However, this parameter has limitations that must be taken into account in the case of impact evaluation. Firstly, the effect identified by this method is entirely dependent on the instrument, which may be affected by treatment choices (Imbens and Angrist 1994). According to Basu et al. (2007), the use of different instruments will produce different effects. Secondly, it is not easy to give an economic interpretation of this metric that identifies impact only in the *completer* subpopulation (Basu et al. 2007; Heckman 2010).

8.4. Results and discussion

8.4.1. *Sociodemographic characteristics of producers*

The sociodemographic characteristics of producers according to adoption status are presented in Table 8.1. This table shows that 92% of the farmers are men and 8% are women. In the subpopulation of adopters and non-adopters, women represent 8.14% and 8.15%, respectively. This reflects the situation in the country where women are the heads of households in the absence of men. The average age of the farmers is 46 years. This average age shows that the active rural population is composed mainly of adults. The average size of a farm is estimated at 18 persons per family. The majority of the sample is made up of producers belonging to the Puular ethnic group (97%). In summary, the results of this table indicate that producers who adopt and do not adopt GAPs differ only in their marital status. In fact, the exposed subpopulation has more married people than the unexposed subpopulation, with a significant difference at the 5% threshold.

Features	**Adopters**	**Non-adopters**	**Total**	**Tests**
Observations	86	135	221	
Age of head of household	47 (13.914)	46 (14.914)	46 (14.399)	Pr (|T| > |t|) = 0.4988
Male (%)	91.86 (0.275)	91.85 (0.274)	91.85 (0.274)	Pr (|Z| < |z|) = 0.9982
Female (%)	8.14 (0.275)	8.15 (0.275)	8.14 (.275)	Pr (|Z| < |z|) = 0.6919
Puular (%)	96.36 (0.018)	97.30 (0.015)	96.83 (0.012)	Pr (|Z| < |z|) = 0.6919
Married (%)	100	95.56 (0.207)	97.28 (0.163)	Pr (|Z| < |z|) = 0.047**
Household size	17 (10.225)	19 (14.541)	18 (13.074)	Pr (|Z| < |z|) = 0.1336

NOTE.– ** denotes significance at the 5% threshold; standard deviations in parentheses.

Table 8.1. *Sociodemographic characteristics of producers (source: authors)*

8.4.2. *Socioeconomic characteristics of adopters and nonadopters*

The statistical analysis of the socioeconomic characteristics of producers is presented in Table 8.2. This indicates that the proportion of adopters who manage to save money at the end of the cashew marketing year is higher than that recorded among non-adopters (53% vs. 25%). It also showed that adopters have easier access to credit than non-adopters. Indeed, approximately 52% of adopters have taken out credit at least once, compared to only 30% of non-adopters. Similarly, approximately 99% of adopters are members of a producer organization, compared to only 27% of non-adopters. We also note that the proportion of non-adopters who have inherited their plantation is higher than that of adopters. Formally, 89% of non-adopters inherited their plantation compared to only 58%. The results of this table also indicate that, on average, the plantations on which GAPs are applied are 7 years old. The difference in age with the non-adopters' plantations, on average, is 4 years and is statistically significant.

Features	Adopters	Non-adopters	Total	Tests
No level (%)	54.65 (0.500)	62.96 (0.485)	59.73 (0.491)	Pr (|Z| < |z|) =0.2211
Primary (%)	29.25 (0.457)	27.41 (0.447)	28.05 (0.450)	Pr (|Z| < |z|) = 0.7897
Literate (%)	16.28 (0.371)	9.63 (0.296)	12.22 (0.328)	Pr (|Z| < |z|) = 0.1424
Access (%)	52.33 (0.525)	30.37 (0.461)	39.91 (0.498)	Pr (|Z| < |z|) = 0.0013***
Savings (%)	53.49 (0.502)	25.19 (0.436)	36.20 (0.481)	Pr (|Z| < |z|) = 0.0000***
OP member (%)	98.84 (0.108)	27.41 (0.448)	55.20 (0.498)	Pr (|Z| < |z|) = 0.0000***
House and plantation (km)	1.43 (1.353)	1.54 (1.871)	1.50 (1.657)	Pr (|T| > |t|) = 0.6339
Plantation heritage (%)	58.14 (0.496)	88.87 (0.315)	76.92 (0.422)	Pr (|Z| < |z|) = 0.0000***
Age of plantation	7 (7.505)	11 (3.025)	9(6.406)	Pr (|T| > |t|) = 0.0001***

NOTE.– *** denotes significance at the 1% threshold; standard deviations in parentheses.

Table 8.2. *Socioeconomic characteristics of producers (source: authors)*

The results in Tables 8.1 and 8.2 reveal a difference in observable characteristics between adopters and non-adopters of GAPs on six variables. These are the variables of marriage, savings, access to credit, membership in a producer organization, land inheritance and age of plantation. The explanation for this difference lies in the fact that most development programs are not randomly distributed among the target population. In fact, individuals who participate in a program generally differ from those who do not. This is because programs are implemented in specific areas and target a particular population.

8.4.3. *The contribution of good practices to performance*

The impact of adopting good cashew agricultural practices on producers' yields is estimated through LATE with and without interaction (Abadie 2003). The results are presented in Table 8.3 and show that the adoption of good cashew nut farming practices has a positive and significant impact on the yield of adopting farmers of 528.84 kg/ha in the model without interaction and 609.87 kg/ha in the model with interaction. These results are consistent with those found by Nirmala (2015), Purnamasari (2017) and Kristin et al. (2010), who showed that farmer training through farmer field schools had a positive impact on crop productivity and per capita farm income. Doukouré (2018) also proves that the project to advisory support through training on good practices had a positive impact on cashew farmers' yields by an average of 141 kg/h. However, in Côte d'Ivoire, Huijsmans (2020) finds that the adoption of GCP does not necessarily lead to a difference in the level of adopters and non-adopters.

Parameters	**LATE- LARF No interaction**	**LATE-LARF Interaction**	**LATE WALD**	**ATE PS**
ATE (kg/ha)				295.25
ATE1	528.84***	609.87***	606.46**	551.35
ATE0	(79.54)	(105.32)	(6872.8)	39.15
PSB				256***
Impact on gender				
Male (kg/ha)	472.77***	537.54***		
Female (kg/ha)	533.56***	615.96***		

NOTE.– *** and ** denote significance at the 1% and 5% thresholds.

Table 8.3. *Impact of the adoption of good agricultural practices on yield (source: authors)*

The coefficient of the average treatment effect in the total population (295.25 kg/ha) based on the matching method (column 4) is lower than the coefficient of the *LATE* estimator (609.87 kg/ha) with interaction (column 2). This result could be explained by the existence of a potential selection bias because treated and untreated individuals are not identical in unobservable characteristics. Therefore, any difference in these results could be partially or entirely attributed to this difference between individuals. Given that the matching method relies solely on observables and completely ignores unobservables as well as the possible endogeneity of the GAP adoption variable, it could be expected to over- or underestimate the true impact of GCP adoption on cashew farmers' performance.

The results in Table 8.3 also show that the impact is much higher for female farmers (615.96 and 533.65 kg/ha) than for male farmers (537.54 and 472.77 kg/ha), regardless of the model with or without interaction. This result can be explained by the fact that women are more respectful of GAPs. Additionally, women are the main actors, along with children, in collection, drying and shelling operations (Sarr 2002).

The determinants of producer performance were estimated from the LARF in its exponential form, and the results are reported in Table 8.4. The joint significance test *F*-statistic (25.67) and the adjusted R^2 show that the model is globally satisfactory. This positive impact of GCP for a number of coefficients of the interaction terms is also statistically significant and thus confirms the heterogeneity of the impact of the adoption of GAPs on cashew farmers' performance.

R-squared = 59.35%

Adj *R*-squared = 56.36%

Wald test for the joint significance of all coefficients (F (3, 177) = 2.62 Prob > F = 0.0426**)

Wald test for non-interacted terms (F (10, 177) = 225.00 Prob > F = 0.0000***)

Performance	Coef.	Std. Err.	T	P>|t|	[95% Conf. Interval]
Married	4.6185180	0.3633892	12.71	0.000	3.901385 5.335651
Age	0.0067517	0.0039335	1.72	0.088	–0.001011 0.014514
Literacy	0.7646304	0.1174846	6.51	0.000	0.532780 0.996481
Primary education	0.2595465	0.4289449	0.61	0.546	–0.586958 1.106051
Household size	–0.0224810	0.0058360	–3.85	0.000	–0.033998 0.010964
Savings	–0.6998352	0.1207025	–5.80	0.000	–0.938036 -0.461634
Access to credit	0.1666327	0.1070197	1.56	0.121	–0.044566 0.377831
Age of plantation	0.0229751	0.0103585	2.22	0.028	0.002533 0.043417
Member of an OP	–0.1114840	0.4329136	–0.26	0.797	–0.965820 0.742852
Distance House orchard	0.0268400	0.0363293	0.74	0.461	–0.044854 0.098534
Annual crop	1.031732	0.6443206	1.60	0.111	–0.239807 2.303271
Breeding adoption	–0.249703	0.0541441	–4.61	0.000	–0.356554 -0.142852
Primary adoption	–0.2841038	0.4386488	–0.65	0.518	–1.149758 0.581551

Table 8.4. *Determinants of producer returns (source: authors)*

Household socio-demographic and economic variables that attempt to explain the positive change in yields induced by GCP adoption include marital status "married", educational level "literacy", age of the producer and age of the plantation. In contrast, other variables negatively influence producer performance. These are household size and the savings variable. The coefficients (4.618 and 0.765) for the marital status variable "married" and the educational level variable "literacy" are positive and significant at the 1% level, indicating that orchards owned by married producers who have been literate yield more than orchards owned by unmarried and illiterate producers. However, the positive effect of the marital status variable is contradictory to the results found by Abbeam et al. (2021) in Ghana, who show that unmarried (single, divorced, widowed) plot managers are technically more efficient than their married counterparts. In a similar vein, Alene and Manyong (2006) also find a negative effect. The coefficients (0.008 and 0.023) of the variables age of the producer and age of the plantation are positive and significant at the 10% and 5% thresholds, respectively, showing that these variables have the potential to increase the returns of adopting producers. In Ghana, Dubbert (2019) finds that an additional year obtained by farming significantly increases cashew yields of non-contract growers but significantly reduces the yield of contract farmers. This result suggests that older growers with older plantings (+10 years)[2] generate more yield than younger growers with younger plantings (–10 years). This result is contradictory to that found by Chirwa (2005), who showed that relatively old farmers are the most reluctant to adopt new technologies. However, the coefficient on the household size variable (–0.022) is negative and significant at the 1% level, implying that household size has a negative influence on performance. This result can be explained by the fact that a good part of the family labor force is not used on the farms. The mechanization of agricultural tools therefore impacts the number of people actually involved in orchards. A similar result was obtained by Ndiaye and Kabou (2021). Finally, the savings variable also has a negative effect (–0.700) on yield.

8.5. Conclusion

Improving agricultural productivity is a recurring theme in development economics. Indeed, increased agricultural growth is often correlated with improved food security and reduced poverty, especially in rural areas. This is why the debate on this issue continues to identify factors that can improve agricultural productivity. We sought to identify the contribution of GAPs to cashew productivity. We argued that adherence to a set of good practices was important if yield was to be significantly improved. Indeed, considering only one factor when studying its contribution to yield can be misleading in agricultural contexts with constraints and

2 Source: Cashew Value Chain Improvement Project Training Manual.

a risky environment. For this reason, we adopted a framework for assessing the impact of adopting a set of GCPs on yield. However, the farmer, based on the information obtained on GAPs, decides to change his behavior by following the recommendations made by the facilitators. From this point of view, the adoption of GAPs is an individual decision generating an endogeneity problem. To better address this problem, we used the instrumental variable approach to estimate without bias the impact of adopting GAPs in cashew cultivation on the performance of producers who were led to change their behavior. The results reveal that the adoption of GAPs allows producers to achieve an additional yield of 609.87 kg/ha. Thus, the extension of these good practices through IRD's field schools in the form of participatory training can make this technology accessible but also increase the yields of adopting farmers. Our results reveal that knowledge through awareness and adoption of sustainable agricultural technologies can help cashew producers to increase their productivity. This implies that when introducing agricultural technologies, it is possible to rely on sustainable practices that do not require drastic costs to use.

From a policy perspective, our results may have important implications for the future of the cashew industry in Senegal. Using a package of GAPs and awareness raising as a vehicle, we found that producers can achieve better returns. Thus, policies to support the cashew industry can be based on respecting minimum spacings of 10 m between trees and 10 m between rows, encouraging producers to prune trees regularly, making them understand the importance of a fence or a firewall to fight against bush fires, sensitizing them to the usefulness of thinnings or weeding the orchard, and associating annual crops with their cashew trees or plowing the orchard.

8.6. References

Abadie, A. (2001). Implementing matching estimators for average treatment effects in Stata. *Stata Journal*, 1(1), 1–18.

Abadie, A. (2003). Semiparametric instrumental variable estimation of treatment response models. *Journal of Econometrics*, 113, 231–263.

Abbeam, G.D., Fusub, S., Ogundejic, A.A. (2021). Technical and resource-use efficiencies of cashew production in Ghana: Implications on achieving sustainable development goals. *Scientific African*, 14, e01003.

Alene, A.D. and Manyong, V.M. (2006). Farmer-to-farmer technology diffusion and yield variation among adopters: The case of improved cowpea in northern Nigeria. *Agricultural Economics*, 35, 203–211.

ANSD (2011). Note d'analyse du commerce extérieur. ANSD, Dakar.

ANSD (2014). Recensement général de la population et de l'habitat, de l'agriculture et de l'elevage. ANSD, Dakar.

ANSD (2020a). Situation économique et sociale du Sénégal 2017–2018. ANSD, Dakar.

ANSD (2020b). Situation économique et sociale de la régionale de Kolda 2017–2018. ANSD, Dakar.

Awio, T., Senthilkumar, K., Christian, O.D., George, W.O. (2022). Yields and yield gaps in lowland rice systems and options to improve smallholder production. *Agronomy*, 12, 552.

Awotide, B.A., Awoyemi, T.T., Diagne, A., Ojehomond, V.T. (2011). Impact of access to subsidized certified improved rice seed on income: Evidence from rice farming households in Nigeria. *International Journal of Sustainable Development*, 2, 12.

Balasimha, D. (1991). Photosynthetic characteristics of cashew trees. *Photosynthetica*, 25, 419–423.

Balasimha, D. and Yadukumar, N. (1992). Effect of plant density on photosynthesis in cashew. *Indian J. Plant Physiol.*, 36(1), 5–7.

Baral, R., Kafle, B.P., Panday, D., Shrestha, J. (2021). Adoption of good agricultural practice to increase yield and profit of ginger farming in Nepal. *Journal of Horticultural Research*, 29(1), 55–66.

Barros, C., Vigil, P., Herrera, E. Arguello, B., Walker, R. (1984). Selection of morphologically abnormal sperm by human cervical mucus. *Archives of Andrology*, 12, 95–107.

Basse, B.W. (2015). Impact de l'adoption des variétés améliorées de riz Sahel sur la pauvreté au Sénégal : approche de l'effet marginal du traitement. PhD Thesis, Université Gaston Berger de Saint Louis.

Basu, K. and Maertens, A. (2017). The pattern and causes of economic growth in India. *Oxford Review of Economic Policy*, 23(2), 143–167.

Cassman, K.G., Dobermann, A., Walters, D.T., Yang, H. (2003). Meeting cereal demand while protecting natural resources and improving environmental quality. *Annu. Rev. Environ. Resour.*, 28, 315–358.

Chirwa, E.W. (2005). Adoption of fertiliser and hybrid seeds by smallholder maize farmers in Southern Malawi. *Development Southern Africa*, 22(1), 1–12.

Crisostomo, B.M.V., Ding, S.Y., Shen, F.X., Sithambaram, S., Gomez, S., Kumar, R., Suib, L.S., Aindow, M. (2005). Synthesis and catalytic activity of cryptomelane-type manganese dioxide nanomaterials produced by a novel solvent-free method. *Chemistry of Materials*, 17(21), 5382–5389.

Deressa, T., Hassan, R., Ringler, C., Alemu, T. (2009). Determinants of farmers' choice of adaptation methods to climate change in the Nile Basin of Ethiopia. *Global Environmental Change*, 19(2), 248–255.

Dieng, F., Ngom, D., Dia, D., Sy, R. (2019). Efficience technique de la production d'anacarde (*Anacardium occidentale*) dans les grandes régions de production du Sénégal. *Int. J. Biol. Chem. Sci.*, 13, 2627–2645.

Dubbert, C. (2019). Participation in contract farming and farm performance: Insights from cashew farmers in Ghana. *Agricultural Economics*, 50, 749–763.

Fe, D.C. and Aklobessi, K.A. (2018). Impact du conseil agricole sur les performances des producteurs d'anacarde de cote d'Ivoire. *European Scientific Journal*, 14(30), 292.

Godtland, M.E., Sadoulet, E., Janvry, D.A. (2004). The impact of farmer field schools on knowledge and productivity: A study of potato farmers in the Peruvian Andes. *Economic Development and Cultural Change*, 53(1), 63–92.

Gosh, S.N. (1989). Effect of nitrogen, phosphorus and potash on flowering duration, yield and shelling percentage of cashew (*Anacardium occidentale* L.). *Indian Cashew J.*, 19, 14–21.

Heckman, J.J. (2010). Building bridges between structural and program evaluation approaches to evaluating policy. *J. Econ. Lit.*, 48, 356–398.

Huijsmans, N. (2020). The potential of good agricultural practices in creating a living income for Ivorian cocoa farmers. MSc Thesis, Wageningen University.

Imbens, G.W. and Angrist, J.D. (1994). Identification and estimation of local average treatment effects. *NBR Technical Working Paper Series*, 118(62), 467–475.

Irz, X., Lin, L., Thirtle, C., Wiggins, S. (2001). Agricultural productivity growth and poverty alleviation. *Dev. Policy Rev.*, 19, 449–466.

Jelsma, I., Lotte, S.W., Jean, O., Arya, H.D. (2019). Do wealthy farmers implement better agricultural practices? An assessment of implementation of good agricultural practices among different types of independent oil palm smallholders in Riau, Indonesia. *Agricultural Systems*, 170, 63–76.

Koffi, S.Y. and Oura, K.R. (2019). Les facteurs de l'adoption de l'anacarde dans le bassin cotonnier de Côte d'Ivoire. *Cah. Agric.*, 28, 8.

McBride, D.W. and Daberkow, G.S. (2003). Information and the adoption of precision farming technologies. *Journal of Agribusiness*, 21(1), 21–38.

Ndiaye, B. and Kabou, A.B. (2021). Adoption des nouvelles technologies rizicoles sur l'efficacité technique au Sénégal. *Science, technologie, développement*, 1(1), 1–14.

Nirmala, G. (2015). Impact of good agricultural practices (GAP) on small farm development: Knowledge and adoption levels of farm women of rainfed areas. *Indian Research Journal of Extension Education*, 15, 153–156.

Nugawela, P., Balde, A., Poublanc, C. (2006). La chaîne de valeurs anacarde au Sénégal, analyse et cadre stratégique d'initiatives pour la croissance de la filière. Programme USAID/Croissance Economique, Senegal.

OIT (2020). *Commerce et chaînes de valeur dans les activités porteuses d'emplois (TRAVERA) : cas de l'anacarde au Bénin*. Organisation internationale du travail, Geneva.

Panco, M., Penot, E., Danthu, P., Michel, I., Jahiel, M., Jagoret, P. (2013). Analyse des savoirs et savoir-faire paysans sur la gestion des systèmes de culture à base de girofle dans le district de Fénérive Est, Madagascar. Working Paper.

Purnamasari, F., Waluyati, L.R., Masyhuri, M. (2017). The effect of good agriculture practices (GAP) on soybean productivity with Cobb-Douglas production function analysis in Kulon Progo Regency. *Agro Ekon.*, 28, 220–236.

Rongead (2013). Connaître et comprendre le marché international de l'anacarde. Report [Online]. Available at: www.rongead.org.

Rubin, D.B. (1974). Estimating causal effects of treatments in randomized and nonrandomized studies. *J. Educ. Psychol.*, 66, 688–701.

Sarr, M.M.B. (2002). Analyse du secteur de l'anacarde : situation actuelle et perspective de développement. Centre du commerce international, CNUCED/OMC (CCI).

9

Bioeconomy and Sustainable Conservation of Plants and Forests in Madagascar

9.1. Introduction

The global challenges of the 21st century are numerous and major: climate change, the rapid scarcity of natural resources, land and ecosystem degradation with a major loss of biodiversity, food and nutritional insecurity, inequalities in access to resources, education and health, and urbanization (Axelos et al. 2020). Bioeconomy is very often put forward as one of the solutions to these challenges. Moreover, Madagascar has an invaluable wealth of pharmacopoeia and traditional medicine. More than 14,000 plant species are listed, belonging to 119 families of Pteridophytes, two families of Gymnosperms and 212 families of Angiosperms. Among these families, five are endemic: *Asteropeiaceae*, *Barbeuiaceae*, *Physenaceae*, *Sarcolaenaceae* and *Sphaerosepalaceae*; and two are quasi-endemic: *Didiereaceae* with seven genera, of which four are endemic to Madagascar, and *Didymelaceae* with one genus endemic to Madagascar and the Comoros.

For more than 80% of the population in developing countries (Stephan et al. 2017), traditional medicine and the use of plants are the first recourse for treatment. In Madagascar, many plant species are thus used as effective remedies to cure various diseases. National consumption is estimated at more than 4,000 tons of fresh

Chapter written by Hery Lisy Tiana RANARIJAONA, Tiantsoa ANDRIANASETRA, Liva Jackson RAHARINAIVO, Vololoniaina RAMAHATAFANDRY, Michaël BEFINOANA, Auguste Botovao RAMIANDRISOA, Cyrille MAHAROMBAKA, Sylvana TOMBOANONA, Chéri Christian TOTONDRABESA, Fenozo ANDRIAMANANTENA, Simon Georges ANDRIANASETRA, Ainazo ANDRIAMANANTENA and Antoine Zafera RABESA.

plants per year (Leipzig 1996). According to Petit et al. (1992), 3,000 species have been identified as having curative virtues, with the rate of use of the species varying from one region to another. However, various anthropic pressures and threats weigh on this invaluable wealth in addition to climate change. However, some plant species are still unknown, especially local populations.

Although individuals' views of botanical gardens vary, they are generally perceived as educational institutions and places of relaxation, where science meets esthetics (Barabé 2012). A botanical garden is defined as "an institution holding documented collections of living plants for scientific research, conservation, display, and education". By this definition, there are over 3,762 botanical institutions worldwide[1]. According to Barabé (2012), botanical gardens offer economic, social and environmental benefits. Economically, they attract tourists and day visitors, whose spending benefits the local economy and the broader community. They are also employers and buyers of goods and services. The collection policy is a fundamental strategic document for all botanic gardens. It guides the development and management of plant collections and helps prevent managers from straying from the intended function and content of collections (Gratzfeld 2020). In addition, gardens enhance the esthetic and recreational value of a neighborhood, increasing the value of neighboring properties and local government revenue from taxes. They are culturally important and reinforce national and local cultural identity.

Today, in some cities of Madagascar, the university is beginning to impose itself not only as a simple service provider but also as an actor which can participate in urban development. For local authorities, the economic impact of public investments needs to be precisely defined. For the universities themselves, a study of their regional positioning can reveal their strengths. The economic contribution of these universities to the development of their host territory can be seen through the direct consequences of the initial investment and the effects of students, staff or operating expenses on the environment (Befinoana and Ranarijaona 2020). In the context of 3,700 botanical gardens worldwide that bring together 60,000 scientists, horticulturists, and specialists in the field of environmental education (Gratzeld, Botanic Gardens Conservation International (BGCI), 2020), the "Vololona Ranjamalala" Educational Botanical Garden (Jardin Botanique Educatif Vololona (JBE)) is a frugal garden which was created in 2018 on the university campus of Mahajanga in northwestern Madagascar with an area of 1 ha (Figure 9.1). This garden fulfills several missions: education, research and ex situ conservation of western Malagasy plants (especially endemic ones threatened according to the International Union for Conservation of Nature) and melliferous plants, nursery

1 Available at: https://tools.bgci.org/garden_search.php.

production of plants from seeds collected for restoration purposes. This garden solves in part the problem of the lack of means for student study trips, the income from the entries being intended to ensure the operation of the garden itself. Consequently, the expenses generated by the PAGESUPRE project must be considered as investment expenses and thus amount to 35,646,132.52 MGA (MGA or ariary (ISO 4217: *MGA*) has been the official monetary unit of the Republic of Madagascar since January 1, 2005), or a value of 14,000 euros in 2018 (Ranarijaona 2019) (8,001.22 euros in 2022). Today, the garden is 4 years old and open to the public, developing societal responsibility and raising awareness of the need to conserve plant species. The garden preserves an endemic palm tree threatened with extinction, and only 30 individuals remain in the wild. Recently, discovered in 2018, *Tahina spectabilis* (Arecaceae) is a plant species that only flowers every 50 years and dies after flowering (Dransfield et al. 2008). Seeds were purchased from villagers at its natural site in *Antsingilava Antsanifera* district, *Sofia* region, (14°45'01 "S, 47°25'53 "E, altitude 9 m). This sale system allows them to keep the remaining specimens. Moreover, the JBE is the only botanical garden to conserve this plant species ex situ in the whole territory of Madagascar. It is connected to the Antsanitia forest (Figure 9.1), known as the *Mandravasarotra* ecotourism site (or "that heals all"), the name given to an endemic plant that is very well known for its virtues and used to prevent Covid-19. It is a dense, dry, deciduous forest in western Malagasy of 47 ha and has degraded because of the illegal exploitation of wood for charcoal and construction wood but also for the illegal exploitation of the *mandravasarotra*. With proper restoration, this forest can be developed into an ecotourism and research site. However, plot 2 of the JBE is equipped with a collection of medicinal plants, including those used by the local population to prevent Covid-19. The health crisis caused by Covid-19 has hit the world since March 2020. In Mahajanga, it has caused significant consequences in the lives of the Malagasy population. A project entitled "the pharmacopoeia of western Malagasy is back for Covid-19: shared knowledge and showcase at the Educational Botanical Garden of the University of Mahajanga" has been conducted in the search for solutions to fight against the Covid-19 pandemic. It consists of updating and improving the pharmacopoeia of the Northwest Malagasy, publishing a booklet on the pharmacopoeia of Covid-19, and creating a database, herbaria, nurseries and living collections "Covid-19" in the Botanical Garden "Vololona" Ambondrona. An inventory and ethnobotanical, sociological and anthropological surveys on Covid-19 and its impact are carried out in Antsanitia and Antrema, with sensitization of villagers to Covid-19 and conservation. In the above-mentioned project, phytochemical and nutritional analyses are carried out for two targeted plants. Knowledge of traditional Malagasy medicine has been inherited from ancestors for several centuries. From a literature perspective, the *Ambongo Boina* pharmacopoeia (Rakotobe et al. 1993) contributes to the Malagasy pharmacopoeias. It contains hand-made plant description plates, vernacular and scientific names of plants with a simplified descriptions of their morphology, types of traditionally known diseases, therapeutic indications and modes of use. However, the *Ambongo Boina* pharmacopoeia is in short supply in libraries and is needed by students, teacher-researchers, and other users. Updating the *Ambongo Boina* pharmacopoeia could make it available and accessible to interested parties and provide knowledge. The updated information from the *Ambongo Boina* pharmacopoeia could serve as a basis for national or international scientists who wish to add value to the plants; in other words, the availability of information is much wider on the Malagasy pharmacopoeia as a result of this research with respect to Covid-19. What is the economic contribution of the UMG to the development of its territory?

The hypothesis that the establishment of a traditional medicinal garden in the JBE is a living heritage of traditional Malagasy medicine to help teach young Malagasy and the entire population to preserve and know the values of nature is put forward.

The objective is to optimize the traditional Malagasy pharmacopoeia for future research. The specific objectives are to (1) multiply the pharmacopoeias of northwestern Malagasy by updating and improving the data; (2) understand the medicinal plants used by the population during this pandemic: curative treatment and preventive fight; (3) conduct ethnobotanical, sociological and anthropological investigations in the 26 fokontany of Mahajanga I and in the degraded forest of Antsanitia to make it possible to analyze the indexes of use of each plant; (4) sensitize the respondents to Covid-19 on the conservation of the forest, which lodges medicinal plants, as the forest is the cradle of life and deserves to be protected for durable conservation; (5) establish a list of plants used, according to the respondents, to treat and prevent against Covid-19 and diseases whose symptoms are close to those of Covid-19, with therapeutic indications and modes of use, to introduce the bioeconomy for a sustainable conservation of the degraded forest of Antsanitia; and (6) enhance researchers by citing the known phytochemical and pharmacological properties of the plant in the updated and improved pharmacopoeia for those plants whose information is already published in scientific journals.

9.2. Study methods

In addition to the bibliographic analysis of the Malagasy pharmacopoeias and Covid-19, the studies were carried out more particularly in the Boeny Region (Figure 9.1), notably in the town of Mahajanga I and in the fokontany Antsanitia.

1) The city of Mahajanga I comprises 26 Fokontany (administrative territorial delimitation comprising sectors or villages) in the district of Mahajanga I, the rural municipality of Mahajanga I. The city is located between 15°39'26.6" and 15°43'21.6" south latitude and 46°18'21.9" and 46°22'07.1" east longitude. The

intervention area is subject to a hot and dry tropical climate. The average annual temperature is 30°C, and the average annual rainfall is 440 mm (ASECNA 2018). The climatic vegetation is characterized by dense dry tropical deciduous forests with a long coastline (DREF Boeny 2012).

2) In the fokontany Antsanitia that hosts the Mandravasarotra forest in Antsanitia, belonging to the University of Mahajanga, in the district of Mahajanga II, the rural community of Belobaka, Boeny region. The forest is located on the northwestern coast of Madagascar, between 15° 33’ to 15° 34’ south latitude and 46° 25’ to 46° 26’ east longitude. It is characterized by its richness in floristic, medicinal and endemic species, which are high but unfortunately heavily exploited for charcoal making and construction.

Ethnobotanical surveys were conducted in July 2020 (the period when Covid-19 was raging in the city of Mahajanga), with 312 respondents (260 respondents in the city of Mahajanga and 52 respondents in the fokontany Antsanitia). The survey is based on the different medicinal plants used by the local population to analyze the use of each plant to prevent the pandemic. To enhance the value of Covid-19-preventive species in the city of Mahajanga I, the questionnaires are based on the identification of plants used to fight against Covid-19-related diseases such as a fever, cough, diarrhea, stomach ache, sore throat and chronic fatigue, as well as their mode of use. Ten people per fokontany were randomly selected for the survey, for a total of 260 respondents. A survey form was prepared with both open and closed questions. The respondents were between 15 and 90 years old.

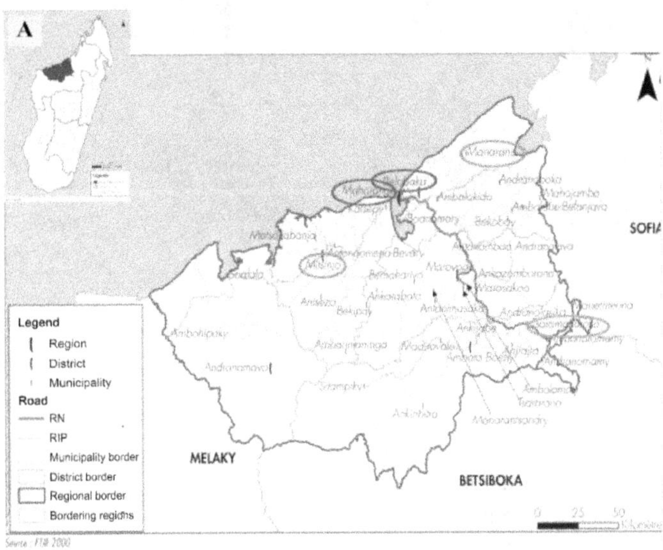

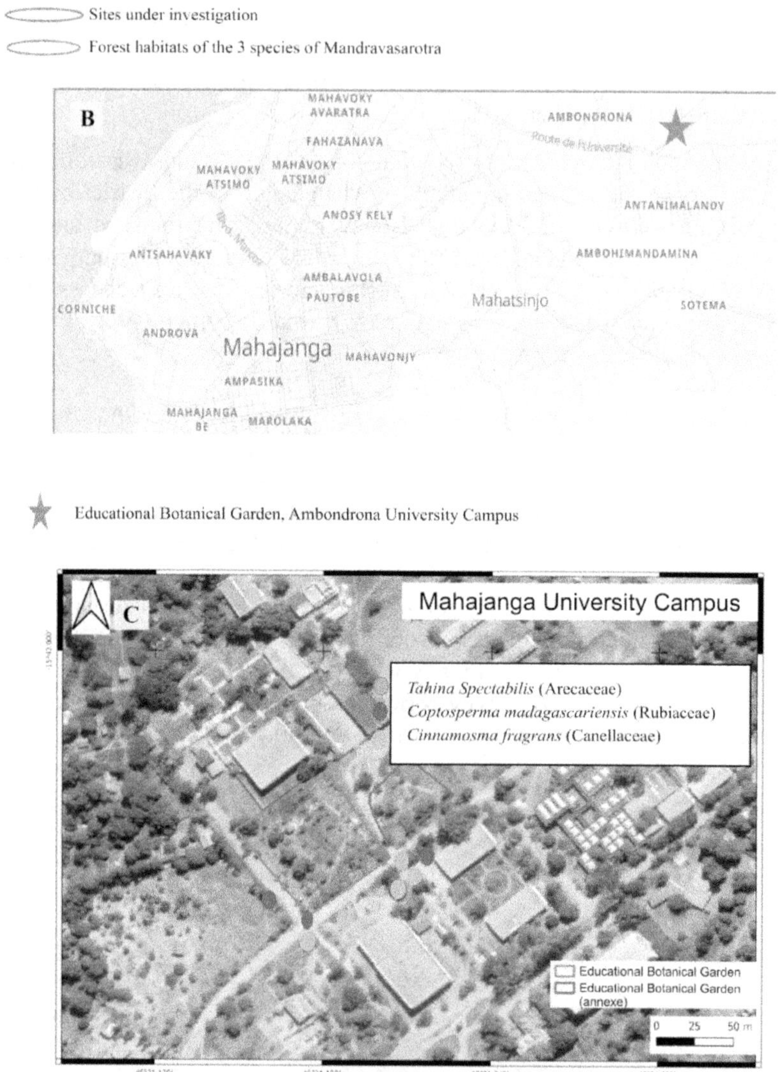

Figure 9.1. *Location of the study sites: (A) Boeny region bounded by the Betsiboka and Sofia regions; (B) the city of Mahajanga; (C) the Educational Botanical Garden. For a color version of this figure, see www.iste.co.uk/casadella/agri-innovations.zip*

Among these respondents, we targeted more traditional practitioners, herbalists, matrons and massagers but also ordinary citizens who have no particular knowledge of plants. The index of use of these target species according to Byg and Balslev (2001) was adopted to express the informant consensus or confirmation index of medicinal plants, with the following formula: CIs = Na/Nt, where CIs is the confirmation index, Na = number of people who cited that species and Nt = total number of interviewees.

In the fokontany Antsanitia, 52 individuals were surveyed in the three sectors, including 22 women and 30 men. To inventory medicinal species at the Antsanitia site, a survey was conducted. Plants with actual or potential economic value, those for which phytochemical and/or pharmacological studies are underway, and those for which ethno-medical data are relevant and worth reporting (Lorre 2006) were then highlighted.

9.3. "Vololona" Educational Botanical Garden (JBE)

Seeds and/or seedlings of plant species of interest against Covid-19-like symptoms (according to the respondents) were collected in the degraded forest of Antsanitia. Seeds (quantity undefined) (Figure 9.2) were dried in the JBE cabin and then placed in nurseries for germination. Ten plants of mandravasarotra and 10 plants of masinjoany were granted (donations) by the private partner "le complexe hôtelier domaine du mandravasarotra", whose headquarters are in the village of Antsanitia, not far from the forest. Flowerbeds have been set up: soil has been ordered and brought by truck to Ambalakida, approximately 20 km from the city of Mahajanga. This potting soil is then transported by the gardeners to the place where the beds are set up. It is added to the JBE substrate, which is very hard and relatively stony. In addition, fertilizers were mixed in to get the beds right. Each bed is covered with "ketihety" or chopped bamboo and held by a raffia thread forming a kind of mat. Plastic bags are purchased for seed placement for certain species whose seeds are not grown directly in the bed. They were nursed in the JBE in November 2020 and collected in the JBE in March 2021.

Abrus precatorius L.

Family: FABACEAE

Common name: Voamaitilany

Vitex bemarivensis

Family: VERBENACEAE

Common name: Mojiro

Rhopalocarpus lucidus Bojer
Family: SPHAEROSEPALACEAE
Common name: Hazondringitra

Commiphora lamii H. Perrier
Family: BURSERACEAE
Common name: Matambelona

Stereospermum euphoroides (Baker) A.DC.
Family: BIGNONIACEAE
Common name: Mangarahara

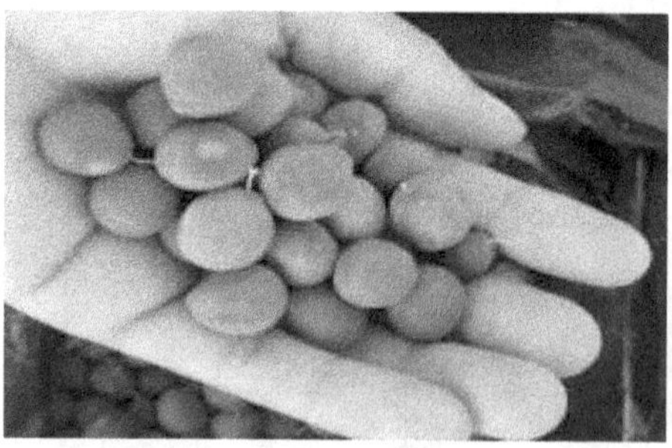

Figure 9.2. *Samples of the seeds of plants used to control Covid-19 collected in the degraded forest of Antsanitia (Ranarijaona et al. 2021). For a color version of this figure, see www.iste.co.uk/casadella/agri-innovations.zip*

9.4. Results

9.4.1. *Perception of Covid-19 in the city of Mahajanga*

A total of 260 individuals were surveyed in the 26 districts of the city of Mahajanga that we visited, including 126 women and 134 men. The categories of respondents ranged from 20 to 40 years old (all genders). All of the people surveyed were aware of the existence of the pandemic in Madagascar and throughout the world. Their mode of acquiring information was radio (53.07% of respondents), television (71.92% of respondents), Facebook (22.69% of respondents) and word of mouth (27.69%). Only three people out of 260 surveyed did not apply any protective measures. Among the plants mentioned by the respondents, "ravitsara" was the plant most used to treat certain symptoms related to Covid-19, such as headaches, flu, coughs and even breathing difficulties. There are several ways to prepare it: either by decoction of the leaves and then using the juice in a drink or bath or for inhalation of its essential oil. It can also be mixed with other plants for drinking herbal teas.

9.4.2. *Perception of Covid-19 in Fokontany Antsanitia*

In total, 52 individuals were surveyed in the three sectors of fokontany Antsanitia that we visited, including 22 women and 30 men. They ranged in age from 15 to 90 years old. All of the people surveyed were aware of the spread of Covid-19. Thus, 45 people (or 87%) of the interviewees had heard information about the Covid-19 pandemic on the radio. To prevent this disease, most of the population in the Antsanitia site preferred to use endemic Malagasy medicinal plants such as mandravasarotra or *Cinnamosma fragrans* Baill. (Canellaceae) instead of applying all the recommended preventive measures for Covid-19. In addition, it is worth noting that despite the villagers' dependence on the Antsanitia forest, overexploitation of wood for construction and charcoal making has been noted locally. Thus, it is necessary to think about the needs of the population in the case of reforestation and/or forest restoration.

9.4.3. *Perception of the use of medicinal plants in the socioeconomic field at the two target sites*

Healing effects: According to the respondents, traditional medicinal plants literally translated as "raok'andro" can relieve pain and cure fevers with flu-like symptoms and neurological signs. The raok'andro can treat any type of illness. With a good traditional practitioner, it can cure even the most serious illnesses.

Economically, raok'andro are affordable: with the purchasing power of everyone, through exploration, they can be sources of currency for the country and sellers.

On the social level, during this Covid-19 pandemic, social cohesion was observed toward the use of traditional medicine: the sharing of plants or leaves, word of mouth advice, and sharing of knowledge on the use of medicinal plants on the radio on various social networks such as Facebook and Twitter.

9.4.4. *Medicinal plants identified*

The medicinal plants identified are (1) those in the literature whose symptoms of use are close to Covid-19 and (2) those according to the responses of the respondents in the two target sites.

In total, according to the census carried out in the literature and the surveys carried out in the target sites, 106 plant species, 26 of which are endemic, treat symptoms, some of which are similar to those of Covid-19. A booklet entitled "*La pharmacopée de l'ouest Malagasy est de retour pour Covid-19 : connaissance partagée et vitrine au jardin botanique éducatif 'vololona' de l'Université de Mahajanga (hay voary_mahajanga)*" was prepared by Ranarijaona et al. (2021) for the description of species, therapeutic uses as well as the modes of use and the known phytochemical and pharmacological properties of certain plants.

9.4.5. *Medicinal plants used to prevent and treat Covid-19-related diseases in Fokontany Antsanitia*

Among the 37 species inventoried, 26 species are medicinal plants (i.e. 70% of the species counted), of which 13 species are endemic. They belong to 24 families and 25 genera, composed of trees, shrubs and lianas. Twenty-two species can treat diseases related to Covid-19, of which eight medicinal plant species (whether from the Antsanitia forest or not) are used by the population of Antsanitia to prevent and treat diseases related to Covid-19: *Cinnamosma fragrans* (mandravasarotra) (*Canellaceae*), *Citrus aurantiifolia* (lemon or voasary makirana) (*Rosaceae*), *Eucalyptus sp.* (kininipotsy) (*Myrtaceae*) and *Zingiber officinale* (ginger or sakaitany) (*Zingiberaceae*) (the latter comes from the highlands). *Cinnamosma fragrans* is the most commonly used species, with an index of use equal to 0.36, followed by lemon and ginger. Fourteen species are known to be wild food plants, of which five species were found during the inventory, and the others were cited by the people surveyed. Among these 14 species, 10 species are fruit plants and four are food tubers.

Impacts		
In the short term	**In the medium term**	**In the long term**
– Knowledge of plants that reduce Covid-19-like symptoms by the public through their display in the UMG Educational Botanical Garden, which will trace the history of Covid-19 in the West of Madagascar.	– Ambondrona Educational Botanical Garden = showcase of traditional medicinal plants including those used for the prevention of Covid-19. – AUF Covid-19 space recognized at regional and national level.	– Space AUF Covid-19 recognized even at the international level: an organized reception that speaks of the cultural, food, cosmetic and medicinal values of plants is planned by the JBE, for international tourists.
– The mandravasarotra, ravintsara, mangarahara (*Stereospermum euphorioides* (Bojer) A. DCBignoniaceae kininimpotsy (*Eucalyptus globulus* (Myrtaceae), oliva or *Eucalyptus citriodora* Hook. (Myrtaceae), can intervene in the treatment against Covid-19. They also provide a justification for the traditional use of aqueous extracts in the form of decoction in the treatment against the Covid-19 virus. – The first three species are endemic.	– Covid-19 can be a tool for conservation awareness: communicating Covid-19 for conservation.	– Further study of the four species.

In the short term (cont'd)	In the medium term (cont'd)	In the long term (cont'd)
– Improved knowledge of the pharmacopoeia. – Improved and reasonable use of plants. – Accessibility of information.	– The leaves of *Ravensara aromatica* (*Lauraceae*) or ravintsara, R. *madagascariensis*, mandravasarotra or *Cinnamosma fragrans*, *Eucalyptus globulus* or kininina, E. *elaeophloia* or olive (Myrtaceae), kininina fotsy or *E. grandis*.	– The populations will also be able to cultivate at home the plants in question in the space, nurseries of the plants are planned.
– A new study space for students (practical work on plant morphology and systematics).	– Plants will be added.	
– Cohesion and solidarity between academic institutions, civil society and the local population.		
– Responsible students and young researchers, know how to share and participate in the actions of the nation.	– Discovery of new plant substances for Covid-19 prevention if grants are available for phytochemical analysis.	– Discovery of new plant substances for Covid-19 prevention if grants are available for pharmacological analysis.
	– Traditional knowledge shared and will be valued by the actors having the means for the well-being of all.	– Visibility of the UMG and Madagascar to the world. – Bioeconomy started.

Table 9.1. *Impacts of creating the JBE annex (Figure 9.1(C)) or plot 2 in the short, medium and long term*

9.4.6. *Medicinal plants to prevent Covid-19 collected in the JBE*

A total of 20 plant species (including six endemics) were available from the nurseries conducted and were collected in JBE plot 2 to prevent Covid-19. Thirteen species are endemic (i.e. 65%), such as *Cinnamosma fragrans*, *Ravensara aromatica*, *Centrosperma madagascariensis*, *Stereospermum europhoroides*, *Caesalpina bonducella*, *Strychnos madagascariencis*, *Dalbergia greveana*, *Salacia madagascariensis*, *Commiphora pervilleana*, *Rhopolocarpus lucidus*, *Vitex beraviensis* var. *acuminata*, *Vitex ankaranensis* and *Mascarenhasia arborescens*.

Table 9.1 shows the short-, medium-, and long-term impacts of creating the JBE annex or Plot 2.

9.4.7. *Suggested plants for the bioeconomy of Mahajanga*

It is important to mobilize more biomass under climatic constraints while preserving the degraded forest of Antsanitia. The endemic plants mentioned above to prevent Covid-19 are suggested as a lever of the bioeconomy and will be able to be the object of restoration in the degraded forest of Antsanitia. We can start with the species for which the demand is already known. The restoration of degraded areas with the mandravasarotra or *Cinnamosma fragrans* and the masinjoany or *Coptosperma madagascariensis* two species very famous for their virtue, the first species is medicinal and the second species is both medicinal and cosmetic.

Both species will be exploitable 4 years after their restoration. With rational exploitation, their restoration could solve the challenge of the preservation of the whole biodiversity space in the forest, the supply of materials and the income of the local populations involved in the first place, the region and the country afterwards. Table 9.2 demonstrates the benefit of biomass valorization for a future bioeconomy.

Current situation of illegal and non-illegal exploitation of trees for charcoal production	**Advantages of the rational exploitation of biomass for a bioeconomy: vision of the future**
One bag of coal costs 15,000 MGA (3.06 euros).	One bag of mandravasarotra leaves costs 150,000 MGA (33.48 euros) to 250,000 MGA (55.81 euros). Based on the data of Andrianoelisoa and Rajoelina: 20,000 kg/year of production of exploitable leaves is equivalent to 3 billion MGA (or 672,105.87 euros) to 5 billion MGA (or 1,120,176.45 euros).
– In 2014: for 1 ha (work of 10 people), valorizable in charcoal only 10 bags on average: sold at 7,000 Mga the bag in the production area, (Razaoharimiasa 2013; Andriamifidy 2014). – In 2022, it was very difficult to find wood for charcoal, often we encountered only the jujube tree. – For 1 ha, only in the Boeny region, five bags can be used as charcoal (personal communication) sold at 50,000 Mga (10.20 euros) after at least 7 days of work for three to five people.	– Restoration with mandravasarotra on a surface of 1 ha: 100 feet *(support of the local populations during the nursery phase of mandravasarotra)*. – One young plant of mandravasarotra, 4 years old, can produce on average: 1 kg of leaves. – For 1 ha (work of five people): 100 feet can produce up to 100 kg of leaves of mandravasarotra, that is to say two bags of leaves, which cost at least 300,000 MGA (that is to say 61.22 euros) per day.

Average number of trees felled per month in an area of 1 ha (average diameter: 5 cm, only jujube trees are currently in the majority in the Boeny region): 9.	– No trees cut down, leaf picking that took place. – On average, 100 m^2 of area saved if no trees are cut down including at least six plant species including animals inside, carbon sequestration achieved, resilience. – Possibility of obtaining the rebate from the economic operator who exploits the leaves for the production of essential oil according to the agreement with the local populations. – Mitigated climate change.
The forest is currently degraded and illegally exploited.	Future development of the degraded Antsanitia forest as an ecotourism research site.
	– Student and faculty research site. – Ecotourism routes enhanced through the joint work of researchers and local populations. – Local people trained for nurseries and as local guides. – Paid admission to the ecotourism site. – Improved bioeconomy of Mahajanga.

Table 9.2. *Interest in biomass exploitation for the bioeconomy planned in the forest of Antsanitia; vision in the future and its valorization in ecotouristic sites of research*

9.5. Discussion

Due to its multivariate ecological factors (altitude, bioclimate, nature of the substrate), Madagascar has particular ecosystems among the richest in the world with an exceptional endemicity rate of nearly 90%: the west (altitude below 800 m) is characterized by a dry dense forest whose leaves are deciduous during the dry season or winter, with many plants belonging to the genus *Dalbergia* and *Commiphora*; the east (with an altitude below 800 m) is characterized by humid dense forests whose leaves remain green and evergreen all year round with plants such as *Anthostema* or belonging to *Myristicaceae*. The highlands (800–1800 m) include humid and subhumid forest formations but with relatively smaller tree sizes and forests characterized by *Weinmannia* and *Tambourissa* species. At the same altitude, but on the west coast, also called western slope wood, a Malagasy endemic forest formation (with a predominance of endemic plant species) is established: it is the forest with tapia or *Uapaca bojeri* and plants belonging to the *Sarcolaenaceae*. Above 1,800 m, we meet the clear forest (trees of relatively small size, even 6 m high) with *Ericaceae*. Finally, Sambirano, located in northwestern Madagascar, has an altitude below 800 m, has a microclimate different from other regions and a particular vegetation with a microendemism. In each phytogeographic distribution, plant species are characteristic, that is, they are found in abundance, and among them are the species of strong use. In addition, each region should have the values of use indexes of plant species, which knowledge is very important; species with high values of use indexes (on the artisanal level, for construction, for traditional medicine and for food, etc.) are those that deserve to be multiplied.

This is the case for the endemic aromatic plant mandravasarotra, which is distributed in the northwest, central-west and east of Madagascar. It is also called motrobeatignana, or saro (vernacular names or appellations vary from one region to another). This genus includes three species of shrubs or trees: *C. fragrans*, distributed in the northern, northwestern and central-western parts of Madagascar; *C. macrocarpa* and *C. madagascariensis,* located in the eastern, southeastern and southern parts of the island (Harinantenaina and Takaoka 2006; Harinantenaina et al. 2007, 2008). Their valorization in their respective regions could start the Malagasy regional bioeconomy through the export of the essential oil extracted from their leaves. Data are available regarding their chemical composition, which is required by sellers in France. They vary from one region to another, and this variation justifies the authenticity of the essential oil. The chemical composition of *C. fragrans* essential oil from two different sites has shown the existence of two different chemical types (Razafimamonjison 2011): linalool for Tsaramandroso (northwest) and 1.8 cineole for Mariarano (northwest), Boeny region (Figure 9.1). In addition, variability in the main constituents of the essential oils of *C. fragrans* is noted. Its essential oil has a broad spectrum of antibacterial activity and could be an alternative to antibiotics in shrimp aquaculture (Randrianarivelo et al. 2009, 2010).

Thus, the species in question has very broad curative virtues and is used in traditional medicine against diseases of the respiratory system, intestinal parasitosis, liver disease, syphilis (Pernet et al. 1957; Schulte et al. 1972) and malaria. Additionally, the essential oil of *C. fragrans* is distinguished by the high contents of neral, geranial and geranic acid, and the essential oil of *C. madagascariensis,* whose major components are caryophyllene oxide and humulene oxide, is also important. The essential oils of these two species are an alternative to antibiotics in shrimp aquaculture (Razafimamonjison 2011). Phytochemical and pharmacological studies carried out by the École Doctorale Génie du Vivant et Modélisation of the University of Mahajanga, currently being published, showed geranic acid, limonene, α-pinene and β-pinene in the majority for the species *C. fragrans* from the district of Mitsinjo, Boeny region, to justify its efficacy in controlling Covid-19 (Roukia et al. 2021). It was the most commonly used species during the Covid-19 period in 2020 according to surveys conducted in Antsanitia and Antrema. In these two forest areas, it is very rare due to overexploitation. It is the same in Mariarano and Tsaramandroso, which truly deserves a large-scale plantation given the current demand. Moreover, an exploitation by collecting the leaves of the local population of Tsaramandroso took place in partnership with scientific researchers within the framework of a project financed by the European Union: GESFORCOM (Community Management, Community, and Local Development). This project has demonstrated the potential of collecting 20,000 kg of leaves per year, 600 kg of oil per year, with an average yield of 3%, and 15–20 kg of oil produced per month.

In Madagascar, 60% of the population is interested in traditional medicine and medicinal plants because of the shortage and high cost of medicines (Leipzig 1996). In addition to the high cost of medicines, the lack of doctors and/or pharmacies encourages people to use medicinal plants to treat themselves, especially in remote areas. Some people simply do not trust modern medicine and prefer to treat themselves in a traditional way.

The number of medicinal plants in Madagascar is approximately 3,245 species belonging to 1,050 genera and 203 families, with an endemicity rate of 60% (Rafidison et al. 2019). However, the exploitation of all these medicinal plants comes from harvesting and collection in the wild. The natural regeneration that ensures production cannot keep up with irrational and uncontrolled harvesting. The use of Malagasy pharmacopoeia has increased during this period of the Covid-19 pandemic, especially concerning curative treatments. Apart from the herbalists who existed in the large markets of the city, a large number of new herbalists have been selling medicinal plants in the streets since the arrival of the pandemic in the city of Mahajanga.

In addition, the JBE garden is a tool for culture, research, innovation, awareness, and conservation for ecological resilience. Its mission is the sustainable conservation and valorization of the collection – the cosmetopoeia and pharmacopoeia ecotourism to promote: research, education, innovation and conservation of various plants (medicinal, cosmetic and endangered), multiplication and seed exchange. Additionally, in the medium term, the garden's mission is to elaborate a connection with the natural forest of Antsanitia, to valorize the latter as an ecotouristic site and to restore it with useful plants with the involvement of the local populations for the improvement of their quality of life. The other missions are the valorization of plants by the extraction of their essential oils, the reforestation and restoration of endemic species and finally the support of research by students from various establishments (Sciences, Tourism, Agronomy, Veterinary, Pharmacy). The JBE was awarded the Francophone responsible innovation label in 2021 by the Agence Universitaire de la Francophonie, thanks to the living collection of plants used to prevent Covid-19. It responds to Sustainable Development Goals 4, 12 and 15 (SDGs, UN, 2015), which are training, research, community service and governance. The JBE is a tool for UMG student practice. It is officially recognized and is beginning to have its place in the city of Mahajanga. In its current state, it is an innovative project for the University of Mahajanga and for Madagascar, as it is the third planting garden in Madagascar after the Botanical and Zoological Park of Tsimbazaza and the Botanical Garden of the University of Antananarivo (Befinoana and Ranarijaona 2020).

The degraded forest of Antsanitia is a forest illegally exploited for the manufacture of coal and the extraction of wood for construction. Its development into an ecotourism research site and its restoration with plants such as mandravasarotra, katrafay and ravintsara with the involvement of the local population could bring a resolution for its conservation. Moreover, the forest of Antsanitia was the natural habitat of the mandravasarotra in large numbers more than a decade ago but has been heavily exploited for its virtue as well as because of the sale of mandravasarotra leaves collected for the manufacture of essential oil. There are only approximately 10 ft left in the area at present. During the spread of this pandemic, the use of this species has become very important, with a use index of approximately 0.36. This plant has several therapeutic indications and can treat several types of diseases, including all the symptoms of Covid-19. However, in nature, the abundance of this species is approximately 0.97%. This situation could therefore endanger the disappearance of this species in this area. Hence, there is interest in restoration with the species in question.

In addition, the interest in restoration requires at least 4 years if we talk about production, that is, the collection of leaves. Another issue is that the low density, seasonality and perishability of biomass represent considerable technological, organizational and economic challenges. To this end, it is necessary to approach the organization of the chains in territories by taking into account socioeconomic expectations. Success and failure, especially with regard to social controversies, the expression of interest groups, forms of opposition and support for bioeconomic projects will have to be analyzed. At the beginning of the project, there is also the need for the availability of the decision support tool through the availability of quantitative and qualitative data (obtaining seeds of the plants to be restored, transport and monitoring, etc.). These data can also be integrated into dynamic socioeconomic models of bioresource management, quantification of mobilized material flows, socioeconomic behavior of actors and production of the industrial apparatus to assess the viability of such transitions (Axelos et al. 2020). The accompaniment of local populations and researchers for the start of the bioeconomy project is necessary, and this is accomplished through responses to a call for research projects and/or entrepreneurship.

9.6. Conclusion

This chapter demonstrates the richness and importance of medicinal plants in the daily life of Malagasy populations. Unfortunately, it is rather a matter of gathering in natural forests, whether protected or not, without any measures of restoration and/or planting of the target species, whether medicinal or not. Such a form of exploitation may not be sustainable. Species with a high utilization index were listed, including those for Covid-19 control. Many of the species are forest species, many are endemic, if we speak only of the species used by the populations in Mahajanga: *mandravasarotra*, *masinjoany* and *mangarahara,* and they could well be used in forest restoration. These are endemic forest plants of the deciduous forest of western Malagasy. The first is a very famous endemic medicinal plant of the west, and the second is a cosmetic endemic plant (powder of the stem as facial mask) with cultural value for the Malagasy women of the west and Comorian. Planting the above-mentioned species would be a very good idea for reforestation given the values of their use index. It is true that Madagascar is very rich in plant species, of which 80–90% are endemic, but the consideration of the requests of the local populations, their needs and the values of the index of use of the species of a given region is very important. The same is true for reforestation and/or forest restoration to achieve ecological resilience and sustainable development and to avoid reforestation failure.

Mandravasarotra or *Cinnamosma fragrans* could be a very productive plant resource for Malagasy populations: its profitability varies from 150,000 MGA (MGA or ariary (ISO 4217: *MGA*) has been the official monetary unit of the Republic of Madagascar since January 1, 2005) (i.e. 33.48 euros) to a minimum of 5 billion MGA (i.e. 1,120,176.45 euros) per year (figure obtained in 2012). However, valorizing without cultivating reduces the income to 0 MGA because at the time we speak, it is very difficult to find leaves of *mandravasarotra,* while obtaining seedlings from nurseries is not easy either. Training and sensitization of villagers and citizens in the city will be necessary to protect together the goods and ecological services provided by biodiversity to humans for future generations. However, beyond that, it is necessary to demonstrate the economic interest of exploitation at the local, regional and national levels, despite the existence of local management transfer in some regions and the impact of exploitation on the conservation of the target species. After 4 years of plantation, leaves of *mandravasarotra* can be collected for sale, and the start of the bioeconomy of Madagascar would be assured with the preservation of this natural resource. Forest restoration or reforestation of this plant could be key to the start of the Malagasy bioeconomy. There are many others, but it is necessary to know the values of use indices in each region because each one has its "emblematic" aromatic plants.

A bioeconomy that serves people, food security and climate change could be created for the "Vololona Ranjamalala" Educational Botanical Garden in the future. A connection to the degraded forest of Antsanitia through the restoration of useful plants such as *mandravasarotra* or *Cinnamosma fragrans*, a very famous endemic plant of the west, is quickly essential. In addition, the valorization of the Antsanitia forest into an ecotourism research site could also increase the financial capacity of the University of Mahajanga but especially for the local populations and those involved in the valorization. With the strengthening of the collaboration of private and public partners, the success of the project toward the bioeconomy of the educational botanical garden and its connection with the degraded forest of *Antsanitia* would be assured and promising.

9.7. References

Axelos, M.A.V., Bamière, L., Colin, F., Dourmad, J.-Y., Duru, M., Gillot, S., Kurek, B., Mathias, J.-D., Méry, J., O'Donohue, M. et al. (2020). Réflexion prospective interdisciplinaire bioéconomie. INRAE synthesis report [Online]. Available at: https://hal.inrae.fr/hal-02866076.

Barabé, D., Cuerrier, A., Quilichini, A. (2012). Les jardins botaniques : entre science et commercialisation. *Natures Sciences Sociétés*, 20, 334–342.

Befinoana, M. and Ranarijaona, H.L.T. (2020). L'Université, économie locale et externalités : le cas de mahajanga. *Revue des sciences, de technologies et de l'environnement (RSTE)*, 2, 298–309.

Byg, A. and Balslev, H. (2001). Diversity and use of palms in Zahamena, eastern Madagascar. *Biodiversity and Conservation*, 10, 951–970.

Convention sur la Diversité Biologique, Biodiversité et Programme de développement durable à l'horizon 2030 (2011). Technical Note, FAO, Rome.

Dransfield, J., Rakotoarinivo, M., Baker, W.J., Bayton, R.P., Fisher, J.B., Horn, W.J., Leroy, B., Metz, X. (2008). Un nouveau genre de palmier coryphoïde de Madagascar. *Journal botanique de la Linnean Society*, 156, 79–91.

Gratzfeld, J. (ed.) (2020). De l'idée à la réalisation – Manuel BGCI sur la planification, l'aménagement et la gestion des jardins botaniques. Botanic Gardens Conservation International, Richmond [Online]. Available at: https://www.bgci.org/resources/bgci-tools-and-resources/bgcis-manual-on-planning-developing-and-managing-botanic-gardens.

Harinantenaina, L. and Takaoka, S. (2006). Cinnafragrins A-C: Dimeric and trimeric drimane sesquiterpenoids from *Cinnamosma fragrans*, and structure revision of capsicodendrin. *J. Nat. Prod.*, 69, 1193–1197.

Harinantenaina, L., Asakawa, Y., Clercq, E. (2007). Cinnamacrins A-C, cinnafragrin D, and cytostatic metabolites with alpha-glucosidase inhibitory activity from *Cinnamosma macrocarpa. J. Nat. Prod.*, 70, 277–282.

Harinantenaina, L., Katsuyoshi, M., Hideaki, O., Masatoshi, K., Kentaro, Y., Yoshinori, A. (2008). Secondary metabolites of *Cinnamosma madagascariensis* and their alphaglucosidase inhibitory properties. *J. Nat. Prod.*, 71, 123–126.

Leipzig (1996). Madagascar : rapport de pays pour la conference technique internationale de la FAO sur les ressources phylogénétiquesi. Report, FAO, Rome.

Lorre, P.I. (2006). Un regard sur l'évolution de la médecine traditionnelle malgache. PhD Thesis, Faculty of Pharmacy, Université Henri Poincaré – Nancy 1.

Moat, J. and Smith, P. (2007). *Atlas of the Vegetation of Madagascar*. Kew Publishing, Kew Royal Botanic Gardens.

Pernet, R. and Meyer, G. (1957). *Pharmacopée de Madagascar*. Institut de Recherche Scientifique Tananarive, Tananarive-Tsimbazaza.

Petit Jean, M., Rakotovao, L., Rasoanaivo, M.P. (1992). *Les plantes utiles de Madagascar : inventaires par genre, espèces et variétés, Tome II*. Académie de Madagascar, Antananarivo.

Rafidison, V., Ratsimandresy, F., Rakotondrajaona, R., Rasamison, V., Rakotoarisoa, M., Rakotondrafara, A., Rakotonandrasana, S.R. (2019). Synthèse et analyse de données sur les inventaires de plantes médicinales de Madagascar. *Ethnobotany Research and Applications*, 18(40), 1–19.

Rakotobe, E., Rasolomanana, C., Randrianasolo, S. (1993). *Pharmacopées de l'Ambongo et du Boina*. CIDST, Antananarivo.

Rakotonandrasana, S.R., Rakotondrafara, A., Rakotondrajaona, R., Rasamison, V., Ratsimbason, M. (2017). Plantes médicinales des formations végétales de la baie de Rigny-Antsiranana à Madagascar. *Revue bois et forêts des tropiques CIRAD*, 331(1), 55–65.

Ramananjanahary, R. and Ratsimiala Ramonta, I. (2005). Études biologique et écologique de deux espèces anti-diarrhéiques d'Ankarafantsika et d'Antrema (cas de *Cabucala erythrocarpa* et *Mascarenhasia arborescens*). In *Recueil de documents pour suivi écologique du programme environnemental*, Roger, E., Rajeriarison, C., Rakouth, B. (eds). Éditeurs scientifiques, Antananarivo.

Ranarijaona, H.L.T. (2018). Jardin Botanique Educatif "Vololona Ranjamalala" au campus universitaire Ambondrona Mahajanga Madagascar. Final report, Projet Fond de Solidarité Prioritaire PAGESUPRE.

Ranarijaona, H.L.T. (2019). Formation en aménagement paysager des doctorants et enseignants de l'Ecole Doctorale Ecosystèmes Naturels (EDEN) : application au Jardin des plantes "Vololona Ranjamalala" Campus Ambondrona. Final report, Projet Fond de Solidarité Prioritaire PAGESUPRE.

Ranarijaona, H.L.T., Voninary, A., Rahilonambinina, F.J., Rakotoarison, E.A., Niembesa, D., Ratsimamanga, A.J.M., Rivonirina, R., Rasolofoniainarinosy, N.M., Tomboanona, S. (2021). La pharmacopée de l'ouest malagasy est de retour pour covid19 : connaissance partagée et vitrine au jardin botanique éducatif "vololona" de l'Université de Mahajanga (hay voary_mahajanga). Report, Livret Covid 19, Projet AUF-COVID 19.

Randrianarivelo, R. (2010). Étude de l'activité antimicrobienne d'une plante endémique de Madagascar *Cinnamosma fragrans*, alternative aux antibiotiques en crevetticulture. PhD Thesis, Faculty of Sciences, Université d'Antananarivo.

Randrianarivelo, R., Sarter, S., Odoux, E., Brat, P., Lebrun, M., Romestand, B., Menut, C., Andrianoelisoa, S., Raherimandimby, M., Danthu, P. (2009). Composition and antimicrobial activity of essential oils of Cinnamosma fragrans. *Food Chem.*, 114(2), 680–684.

Razafimamonjison, D.E.N.G. (2011). Variabilités chimiques et activités antimicrobiennes des huiles essentielles de Cinnamosma spp. (alternative aux antibiotiques en aquaculture de crevette). PHD Thesis, École Supérieure des Sciences Agronomiques, Université d'Antananarivo.

Roukia, D., Razafinirina, E.I., Omary, H., Randriamaharosa, B., Temasoa, Y., Rajaonarison, J.F. (2021). La pharmacopée de l'ouest malagasy est de retour pour covid19 : connaissance partagée et vitrine au jardin botanique éducatif "vololona" de l'Université de Mahajanga (hay voary_mahajanga). Analyses phytochimique et nutritionnelle. Report, Projet AUF-COVID 19.

Schulte, V.K.E., Rücker, G., Lewe, W.A. (1972). Über Die in altsstoffe von *Cinnamosma fragrans* Baillon-einer arzneipflanze aus Madagaskar. *Deutsche Apotheker Zeitung*, 41, 1578–1579.

Wyse Jackson, P.S. and Sutherland, L.A. (2000). *Agenda international pour la conservation dans les jardins botaniques*. Botanic Gardens Conservation International, Kew.

10

Bricolage in Agriculture Sector: Emergence Dynamic and Consequences in Vietnam

10.1. Introduction

Agriculture is always a critical sector of several countries around the world, regardless of whether they are industrial, service or agricultural nations. The importance of the agriculture sector has long been studied in the world of economics and management science. In developed countries, the agriculture sector is strongly developed and industrialized due to early and large industrial evolutions. However, agriculture in most developing and less-developed countries is predominantly backward and resource scarce. Farmers in such nations, mainly small-scale production, lack numerous basic resources, such as financial resources, infrastructure, human resources, knowledge, and technology. Operating under such a resource-constrained context, agriculture in developing and less-developed nations has applied several simple and traditional methods, even sometimes unconventional ones, in growing plants or raising livestock.

In the literature on management, bricolage first emerged as a construct that describes how individuals improvise "by recombining existing, but individually less useful, resources to create value through creative reconstruction". Bricolage has also been introduced as an innovation approach to achieve good-enough outputs with a limited amount of resources (Baker and Nelson 2005). Among the initiatives in agriculture, bricolage came up frequently in the discussion on how farmers can overcome their inherent resource constraints. For example, instead of purchasing, designing and developing new equipment that seemed out of reach for resource-limited

---

Chapter written by Son Thi Kim LE, To The NGUYEN and Phuong Anh Thi NGUYEN.

farmers, bricolage was introduced in some African countries as a good-enough approach to tinker or adapt drip irrigation systems (Benouniche et al. 2014) and agricultural equipment (Sboui et al. 2022) for the specific needs of farmers. At a larger scale, bricolage behavior also exhibits benefits in that farmers cope with the challenge of other resource constraints (Tindiwensi et al. 2020). For example, how the farmers innovatively reallocated and effectively exploited their existing resources such as land, finance and labor that seem to be inexpensively available or easily accessible and attainable.

Regarding a common point between the bricolage and agriculture sectors – environmental issues – in one aspect, the agricultural sector causes several negative effects on the external environment (e.g. soil erosion, pollution of the seas, rivers and lakes, groundwater, chemical pollution of the top layer of the earth, and extinction of some species of animals and plants) (Tsoraeva et al. 2020). However, in another facet, bricolage has recently been studied as an innovation approach using limited resources and allowing innovation toward environmental sustainability and frugality to be developed, thus contributing to overcoming environmental issues (e.g. Vilchez and de la Hiz 2018; Musona et al. 2020; Wu and Liu 2022). As a result, the emergence and development of bricolage in the agriculture sector is expected to overcome the resource-constrained situation of farmers and simultaneously limit the environmental issues caused by agriculture production activities.

Among developing countries whose agriculture sector is dominant, Vietnam appears as a typical example to expose and explore how bricolage has emerged and supported the limitation of the weaknesses of agriculture production activities, including resource-poor settings and environmental deterioration. Vietnam is an agricultural nation with a large proportion of its population working in the agricultural sector, for instance, approximately 62.5% in 2000 and 36.23% in 2020 (World Bank 2022). The portion of employment by agriculture is still more than double the portion of agriculture in GDP (13.96% and 12.36% in 2019 and 2021, respectively). These data show the relatively low productivity of this sector. Vietnam is still one of the world's leading rice producers and the world's second largest exporter of rice, and the Vietnamese agriculture sector accounts for 13.2% of exports (World Bank 2022). However, the environmental issues induced by the agriculture sector in Vietnam are also serious: agricultural production activities accounted for approximately 33% of total GHG emissions in 2010, just after the manufacturing sector, in which rice cultivation and livestock production are the major sources of GHG emissions.

The objective of this chapter is to explore how bricolage can emerge and develop in the scarcity context of the agriculture sector, and its possible role in liming environmental issues due to agricultural production activities. First, the definition and environmental domains stimulating bricolage and the emergence of bricolage in agriculture sectors among several countries will be summarized, especially in developing countries. The next section focuses on the emergence of bricolage when Vietnamese farms cope with each resource constraint, and then the potential of bricolage in solving the environmental problems in Vietnam.

10.2. Bricolage in the agriculture sector

10.2.1. *Bricolage and scarcity*

The term bricolage originated from the work of French anthropologist Claude Lévi-Strauss, in which bricoleurs are described as those who found solutions for their problems by relying exclusively on "whatever is at hand" (Lévi-Strauss 1967). Resources "at hand" include personal knowledge and competence, initial investment and personal networks (Mahajan 2013). When trying to explain bricolage, authors have paid specific attention to scarcity insofar as it appears as a constitutive condition. Indeed, in cases where resources are abundant, people have little interest in engaging in bricolage. In contrast, under scarcity, people are more inclined to use items at hand in novel ways that were not originally intended (Rosenzweig et al. 2016). For that reason, scarcity status has been studied as a trigger of bricolage (Garud and Karnøe 2003; Halme et al. 2012).

The question is how bricolage emerged from the use of all existing resources, regardless of whether they are useless or substandard, to create "brilliant unforeseen results" (Lévi-Strauss 1967, p. 17). Bricolage thus constitutes another promising lens through which to examine how resources are assembled and gathered in resource-constrained settings. Common points of bricolage at the individual, organizational and interorganizational levels are "bias for action", "an active engagement with problems and opportunities", "the capability of making do, using and (re)combining resources at hand" and "relaxing the rules of what resources could or could be used for" to create good-enough solutions (Baker 2007).

Specifically, faced with challenges where the response seems to be too resource intensive, people tend to wait for the necessary resources they require. However, bricoleurs follow the traditional guideline that "necessity is the mother of invention" (Senyard et al. 2014) and thus will gladly tinker with all existing resources to find some kind of response. In doing so, they show a bias toward action and active engagement with problems or opportunities (Baker and Nelson 2005). They tend to ignore the limitations of commonly accepted definitions of material inputs, practice and definitions and standards, and instead emphasize trying solutions, observing and processing the results. Through reconstruction, recombination and reuse of all resources at hand for novel meaning, purpose and value (Salunke et al. 2013) that could be far from the original one of such resources, the bricoleurs show their creativity more than non-bricoleurs (Senyard et al. 2014). In summary, it is worth noting that bricoleurs do not only compensate for the lack of resources. Because bricolage involves creativity, not only does it help create new from old, but it also generates *something from nothing* (Baker and Nelson 2005; Cunha et al. 2014).

10.2.2. *Three main domains stimulate bricolage*

To better understand how bricolage emerges, it is necessary to explore how the resources at hand are used to cope with resource shortages. Resources at hand are any means available that can help bricoleurs, such as material inputs, knowledge, information, experience, technologies, free time, private and professional networks, etc. Bricoleurs were characterized by "the collection and storage of physical inputs, origin and breadth of the skills used by personnel, nonconformity to craft standards, industry practices, and legal regulations, and the nature of their social networks" (Baker and Nelson 2005, p. 344). Baker and Nelson (2005) thus note that bricolage may manifest in three main domains, including inputs (e.g. materials, skills, labor), the regulatory and institutional environment, and customers and markets.

Inputs

The inputs include, as proposed by Baker and Nelson (2005), physical resources and labor and skill. Regarding physical inputs, bricoleurs are more likely to collect, acquire and extract usage from resources that others reject or do not intend to use. As a result, compared to non-bricoleurs, bricoleurs tend to maintain a diverse collection of materials, tools, parts, equipment and other physical inputs (Ronkko et al. 2014). Bricoleurs are those who can "make do" with the cheap and free resources available (Baker and Nelson 2005). Bricoleurs often collect and keep resources "just in case" they need them in the future (Lévi-Strauss 1967). Regarding labor resources and skills, Lévi-Strauss (1967) introduced the concept of bricolage upon the comparison between engineers and bricoleurs. Accordingly, the engineer, who has received expert education within a well-defined field of knowledge, intensely understands the structural characteristics of things for which general rules can be adopted and has predetermined ways of using their resources. When facing a new task beyond their expert field of knowledge, the engineer is more likely to improve the needed skills through training or avoid the task by handing it over to better qualified experts (Duymedjian and Rüling 2010). However, bricoleurs are willing to "make use" of a variety of amateur skills and resources available (Baker and Nelson 2005), without worrying too much about whether they have the right tools and resources at hand (Senyard et al. 2014). Bricoleurs, therefore, are often people with a variety of knowledge, skills and experiences that are self-learned and informally picked up, rather than formal and intensive training as engineers. These broad self-taught skills are "applied with little regard to craft or professional boundaries" (Baker and Nelson 2005, p. 345). The core of the bricolage framework, in this case, the bricolage domain, is resource combination and transformation from scarce and simple inputs at hand into useful outcomes (Desa 2012).

Regulatory and institutional environment

During the process of creating possible solutions with available resources, the activities of bricoleurs may often extend beyond the limit of local regulations and professional and industry standards and norms (Baker and Nelson 2005). For example, their activities can ignore the environmental and work safety rules (Ronkko et al. 2014), or even "break the law" by operating without permission or not meeting the obligatory requests of the authorities. Instead of complying with the rules and regulations, bricoleurs thus have explored the extent to which external rules and standards place real constraints on them. In doing so, over time, they have accumulated a comprehensive understanding of what they can "get away with" and what constraints they can ignore and even disregard (Baker and Nelson 2005).

Customers and markets

Customer and markets could be a trigger for bricolage emerging as a resource at the hand of bricoleurs. For instance, Baker and Nelson (2005) observed an interesting phenomenon in the network ties of bricoleurs. Precisely, the bricoleurs, instead of sustaining a professional distance from their customers, employ their customers as a resource – often for labor and expertise (Ronkko et al. 2014), because the customers could have an understanding about the products/services that they want to buy. Similarly, suppliers have also become customers and vice versa. Bricoleurs along with their customers also identify scavenging opportunities together and sometimes act as brokers for each other for the purchase of physical inputs. The informal and multiplex relations create a community of friends where they can make use of scavenged resources and self-taught skills transferred among network members (Baker and Nelson 2005).

10.3. Bricolage in agriculture sectors – How does it work?

As mentioned above, agriculture is generally known as an important sector but is still underdeveloped in several countries, especially developing countries. The backward and fragile condition of agriculture in such countries forced farmers to discover and develop unconventional methods that were suitable for the specific needs and current conditions of farmers in the farming process. To explore how bricolage works in agriculture, this chapter analyzes the emergence of bricolage processes in combining two axes: (1) typical actions of bricoleurs (such as refusing to enact multiple limitations; making do; recombining resources "at hand"; relaxing the rules of what resources should or could be used for; etc.); (2) three domains of bricolage (physical inputs, labor and skills; regulatory and institutional environment; and customers and markets).

One typical example of bricolage in agriculture in Africa is its application to the construction and improvement of drip irrigation systems. In particular, Benouniche et al. (2014) presented the main problems of Moroccan farmers in constructing and using drip irrigation systems based on international standards. They lacked financial resources to invest in it, met several other technical problems during the use process and lacked experience and knowledge about using this system. To cope with such scarcity, bricoleur farmers modified and adapted the standard design to generate their own technical standards. This process of bricoleur farmers in Morocco developed over time through three main periods. In the first period, international standard equipment – drip irrigation systems – were imported and installed at large-scale farms. During the process that farm laborers helped to install this system, they gradually discovered and learned about new techniques. "They discussed with the engineers and technicians who came to install the systems and observed. They were curious about the new technique, the different components of the system, and their use" (ibid., p. 653). In doing so, the interactions with their suppliers put them in the process of accumulating self-taught knowledge, experience and skills on the operating system. Furthermore, this self-study process was developed during the use process due to some difficulties, such as lack of training, experience and local support on how to operate and maintain the systems, as well as long geographical proximity with suppliers. Refusing to enact multiple of the above limitations, all these factors encourage farm laborers to engage in "making do by relying on their own ingenuity" to seek out possible solutions and then indirectly enrich their experience and skill when using drip irrigation systems. In the second period, the farm managers and laborers, day by day, know how to deconstruct drip irrigation systems and design alternative low-cost systems. Particularly, going beyond the operation and maintenance of this system, bricoleur farmers start to physically modify their own systems. They "gradually adapted the system to their personal situation (knowledge, training) and to the farm structure (cropping systems, topography, who is irrigating). They removed certain components, and replaced others that regularly broke down, or did not function satisfactorily" (ibid., p. 653). In particular, the farmers who installed this modified system encountered problems with the filter systems and found a way to replace them with local low-cost alternatives by making use of local support sectors (such as advisers, fitters, retailers and welders). Day by day, the bricoleur farmers, regardless of farm scale, started to adapt the drip irrigation systems to best suit their physical and socioeconomic constraints. At the end of this process, "three alternative types of drip irrigation systems progressively became available on the local market, installed by local fitters or farmers themselves" (ibid., p. 654). To compensate for their shortage of infrastructure – a production factor – bricoleur-famers employed their self-study experience and skill accumulated during the first period to make do and recombine all local resources available to create a physical input that was personalized for their own situation. Could the new and personalized system that was created by modifying an imported international standard system be officially approved by the authorities? During the second period, it is clear that the local drip irrigation systems were tailored to cater only to different physical and socioeconomic requirements without regard to compliance with standards. Fortunately, however, this process resulted in a localization system that gradually became the local standard.

Similarly, a study by Tindiwensi et al. (2020) on commercial smallholder farms showed a model of entrepreneurial bricolage that is exhibited in different ways and dimensions when coping with the constraints imposed by resource environments. In particular, small firms could "create something from nothing" in resource-constrained environments when they refuse to enact the limitations imposed by resource environments. They creatively reallocate their existing family resources (including land, income and labor) and use existing material and farm waste for new use to extend beyond the limits of the farm resource environment. For instance, first, to address the lack of financial resources, smallholder farms obtain cheap financial resources by (1) getting interest free "loans" from their family and relatives (Anderson et al. 2016) and (2) having multiple income sources and flexibility in using income from other sources for farming activities. The latter means that the farmers who may have several sources of income from other small business activities, in some cases if necessary, reallocated their financial resources for farm activities depending on priority and vice versa. At the same time, when budgeting their farming activities, they plan their expenditure based on the budget they can have and avoid spending or borrowing for unexpected expenditures. Second, family land could be used by small farmers to start their farming businesses. Third, to cope with a lack of labor sources, small farmers heavily rely on family labor, with some family members involved in farming activities. Furthermore, to attain more technical and farm management skills, small farmers rely on informal and experiential learning processes. Contact with extension workers from the government, non-governmental organizations (NGOs) and free training programs provides small farmers with more agricultural and farm management skills and knowledge. However, studies in other African countries have also indicated that collaborating with extension services is not truly strong and sometimes does not allow timely advice for farmers (Tarchiani et al. 2017). Last, as members of informal farmer groups, social relations or networks with other farmers allow small farms to collect market information, as well as technical and management skills and experiences about the farming process. The multiple actors in smallholder sectors (Kilelu et al. 2017) enable each farm member to obtain more access to training and input from stakeholders (Faure et al. 2017).

The two above examples of emerging bricolage in different environmental domains show numerous consequences of bricolage. Bricolage, which tends toward frugality, primarily enables farmers to overcome scarcity contexts and then limit environmental issues by: making do and recombining all resources available, including standard and non-standard material inputs, using and recycling farm waste for new purposes, exploiting local low-cost resources to replace high-standard resources as much as possible, finding ways to use available resources in a non-wasteful manner, etc. All these outcomes of bricolage in farming activities are consistent with several prior studies about the potential of bricolage in reducing environmental issues by combining waste and other resources at hand to generate new green products (Musona et al. 2020), using the by-products of one plant as an input to another to reduce resource consumption (Vilchez and de la Hiz 2018), recycling both fiber waste and precipitated sludge and reducing the use of raw materials (Wu and Liu 2022).

10.4. Bricolage to improve Vietnamese agriculture in a scarcity context

The agricultural sector of Vietnam has experienced remarkable growth over the last few decades, with an annual growth rate of 3.5% during the period 1986–2017. However, the whole country has 36% of the population engaged in agriculture, but this sector contributed a very small part of GDP in 2020. The share of agriculture in GPD is forecasted to decline over the coming two decades, by approximately 0.5% per year. Primary agriculture is expected to account for some 8–9% of Vietnam's GDP by the early 2030s. The reasons for this reality were pointed out by the Vietnamese government: (1) agriculture is backward, slow-growing, spontaneous, fragmented, small-scale (average 0.18 ha/land plot and 2.5 plots/household), low output and poor quality compared to other countries in the region; (2) labor productivity in this field, as well as productivity growth is still low; (3) investment in agriculture is quite low, with only approximately 1% of the total number of enterprises (mainly small and micro enterprises) having investment in agricultural production; (4) capital mobilization, access to credit and banking services are limited, capital costs are still high, approximately half of rural households have difficulty accessing banking services, and informal credit ("black credit") is still present in many rural areas; (5) the application of advanced technology in agriculture is generally weak, with outdated technology, leading to the fact that 90% of Vietnamese agricultural products are exported as raw materials or unprocessed products; and (6) the income of several rural people is still very low.[1] Under that resource environment, how does bricolage emerge when farmers cope with a lack of resources, and what are the consequences of bricolage in Vietnam?

---

1 See: http://nongnghiepnhatban.com.vn/nong-nghiep-viet-nam-con-lac-hau-cham-hon-nong-nghiep-mot-so-nuoc-trong-khu-vuc-T32d151v3056.htm.

10.5. Infrastructure resources and advanced technology

Infrastructure and technology are among the primary conditions for agricultural development and livelihood improvement. However, Vietnamese farmers often experience fragmented infrastructure and backward technology in their farming activities. Through bricolage, several Vietnamese farmers tended to use and make-do with all resources that they have and can access to in order to design and construct their own physical machines and equipment. Their results could not only be simple solutions based on their traditional experiences using possible local low-cost alternatives, but also a complete machine or equipment that does not meet high standards in terms of performance and technical standards, but is more suitable for their needs in terms of its functions, technology and expenditure.

For instance, in some fishery farming areas of Vietnam, the infrastructure is also poorly developed. The construction of fish or shrimp farming is often from many years ago, and the irrigation system consists of many long, narrow canals meandering through the village, causing water pollution, water shortages and risk of transmission of disease (Ha et al. 2013). In addition, the average farm size was relatively small, so farmers had no room for water treatment. Without sufficient financial resources, many farmers approached their infrastructure problems in a traditional manner developed through their own experiences. For instance, in Ho Chi Minh city, a farmer named Pham Thanh Tinh uses a large canvas around the shrimp pond to make the water cleaner so that toxic impurities accumulated in the ground do not penetrate the shrimp causing disease. In the past, Mr. Tinh raised shrimp in earthen ponds and had to expend a lot of effort to renovate and clean the land carefully. With the canvas ponds, the cleaning work takes place very quickly, draining the water and then releasing the water for shrimp farming right away.

In agriculture, farmers tend to be exposed to harmful chemicals in farming work. Perceiving the potential risk for the traditional method of farmers, Le Van Sua (Dong Thap, Vietnam) hopes to develop an advanced technology system such as an eco-sprayer so that farmers will not have to be exposed to harmful chemicals. As a result, he stopped farming and switched to learning mechanics for two years. After that, he started making small assemblies and putting parts together to operate. For each successful machine, he carefully recorded the feedback from the farmers who had used the machine and adjusted each part and each terrain. In the countryside, with insufficient equipment and materials, he started looking for recycled parts in local stores. From the conventional sprayer, Nguyen Van Sua has improved and invented new sprayers for paddy rice called the three-in-one fertilizer machine. Farmers using his machine will save on hiring manual labor for spraying, fertilizing and sowing rice.

Figure 10.1. *Pham Thanh Tinh checks shrimp food in the canvas (source: https://tepbac.com/tin-tuc/full/cach-nuoi-tom-sang-tao-trong-ao-lot-bat-21642.html). For a color version of this figure, see www.iste.co.uk/casadella/agri-innovations.zip*

Figure 10.2. *Le Van Sua and his multipurpose sprayer (source: https://vietnamnet.vn/nong-dan-lop-9-che-may-da-chuc-nang-xuat-khau-sang-campuchia-367426.html). For a color version of this figure, see www.iste.co.uk/casadella/agri-innovations.zip*

Similarly, Ta Dinh Huy (in Hanoi) is one of the 10 typical young faces of the country with diligent manufacturing and improvement of agricultural machines. As a matter of fact, he often sees farmers working hard all year round in the fields, but with low labor productivity. Thus, he decided to start with the agricultural machine. His latest machine has 15 functions, including plowing, creating beds, milling the soil, harrowing, pumping water, creating rows for sowing seeds, weeding the garden, pulling trailers, winches, digging coffee tubs, turning biofertilizers, spraying pesticides protecting plants, generating electricity, grinding animal feed and digging holes for planting trees.

Figure 10.3. *Ta Dinh Huy and his multifunction machine (source: https://tuoitre.vn/nong-dan-tre-che-tao-may-nong-nghiep-15-chuc-nang-1282937.htm). For a color version of this figure, see www.iste.co.uk/casadella/agri-innovations.zip*

Pham Van Hat, a farmer in Hai Duong, has also invented 40 kinds of machines for agriculture, forestry, handicrafts and craft villages. Among such machines, the invention of a robot to place seeds, insecticide sprayers, and potting machines has been exported to many countries. The price of his products is also very competitive. The most expensive product compared to foreign machines is only one-tenth cheaper, but can replace the labor of 40 workers.

Figure 10.4. *Pham Van Hat and his seeding machine (source: https://baodansinh.vn/nga-mu-vi-anh-nong-dan-sang-che-may-nong-nghiep-xuat-khau-di-nhat-my-20191103124741472.htm). For a color version of this figure, see www.iste.co.uk/casadella/agri-innovations.zip*

Examples of bricolage in manufacturing and improving agricultural machines abound among Vietnamese farmers. Bricolage is here manifested by their refusal to be constrained by the limits (in terms of material inputs, technology, knowledge, skills, finance sources) to make do, recombine and exploit all of the things they have or can access to create their own physical inputs without worrying about the rules of what resources should or can be used for. These machines and equipment, originally developed for the specific needs of farmers, were therefore not traded in the market due to failure to meet regulatory requirements. However, through the exchange of information with farmers-customers about their needs, the advantages and disadvantages of using these machines, and the individualization requirement for the situation of each farmer-customer, such machines have been perfected over time and can finally be tradable in both domestic and international markets. This confirms the role of the institutional and regulatory environment and customers and markets as contextual factors that stimulate bricolage activity.

Another popular solution for infrastructure shortages in rural areas of Vietnam is the integrated production models of crop/livestock farms. This approach helps to reduce the amount of agricultural waste that is usually discharged into ponds or streets without appropriate treatments. This approach has become increasingly popular in the economic development of the Vietnamese rural sector. For instance, agriculture (post-harvesting) and livestock farm wastes were collected for biogas production and then supplied to living activities and biofuel production; sludge generated from these processes was used as fertilizer for crops. This solution shows the emergence of bricolage through recycling and transferring farm waste in livestock production to become an agriculture production factor that is inexpensively available. In doing so, bricolage solutions partly limit the environmental issues caused by agricultural activities.

10.6. Financial resources

One of the foremost scarce resources of Vietnamese farmers is financial resources to invest and operate in normal conditions (Minh et al. 2010). To cope with these situations, several methods have been introduced, and the following are the best.

First, farmers frequently borrow from relatives or friends at no charge or low interest rates. However, they cannot borrow a large amount of money from relatives and friends because they usually deposit their properties at banks to gain monthly savings interest.

Second, farmers can borrow from commercial banks or other microfinancial institutions. However, collateral is needed, and farmers can only offer their land use certificate (LUC) with a low value specified by the government. Accordingly, farmers usually have insufficient collateral, which prevents them from borrowing capital from those institutions.

Third, farmers use the group-lending technique by joining production chains or local mass organizations, for example, women's or farmer's unions, to obtain union loans through the Vietnam Bank for Social Policies. In June 2015, the Vietnamese government issued Decree No. 55/2015/ND-CP on credit policies for agricultural and rural development to encourage agricultural production under the forms of associations, cooperatives and unions, which can lend up to 70% of the value of the project without collateral. This method not only creates more resources, but also raises the responsibility of farmers in production or unions. To apply this technique, such unions need the approval of the local authorities to operate, which is equivalent to political or administrative organizations (Duong and Izumida 2002). In this case, group members are not liable for the debts of other members. This procedure relieves the lender from investigating the creditworthiness of individual borrowers but the union. Regarding production chain loans, this technique applies for households (cultivating input materials), factories and cooperatives (producing and processing agricultural products) within the value chain of agricultural products. In the Ninh Thuan province, many production chains, such as growing grapes, sugarcane, cassava, and medicinal plants, have been promoted in recent years. As a result, several commercial banks, such as Lien-Viet-Post Bank, have reached out to many grape farmers along the value chain in the province to provide them with loans.

Figure 10.5. *Lien-Viet-Post Bank officers in rubber tree cooperatives (source: https://thanhnien.vn/cho-vay-chuoi-gia-tri-thuc-day-phat-trien-kinh-te-nong-thon-ben-vungpost1451144.html). For a color version of this figure, see www.iste.co.uk/casadella/agri-innovations.zip*

Fourth, instead of joining group lending, farmers can participate in cooperatives and borrow internal credit. This method has been developing rapidly in recent years thanks to its ability to meet capital for farmers more quickly than banks and other credit institutions. Currently, Vietnam has nearly 1,200 agricultural cooperatives carrying internal credit, equal to 10% of the country's total number of agricultural cooperatives. The amount of capital mobilized by cooperatives varies significantly from a few hundred million to several billion dong/cooperative. The procedure is simple but still ensures the ability to recover, so both parties benefit and have safety assurance through the intermediary role of the cooperative.

Fifth, farmers can join the conventional financial model called "Hui". "Hui" is an ancient form of quick capital mobilization in Vietnam without going through many procedures and approval steps, such as a bank loan or cooperative loans. It aims for the mutual assistance of citizens and shall be implemented under law. "Hui" means a form of transaction regarding property following customary practice based on an agreement reached by a group of people who assemble and jointly determine the number of people, the term, the amount of money or other property, the form of contribution and payment of "Hui", and the rights and obligations of the members of the group (Vietnamese Civil Code 2015)[2]. After receiving the "Hui" money to use, the participants continue to pay until the end of the cycle. The last person in the round has been "hooked", and all participants must obey the rules of "Hui". In the past, this method was considered illegal in Vietnam, but was legalized in 2015. However, this capital raising form contains many potential risks that farmers must carefully consider before participating.

In addition, farmers diversified their income by doing non-farm jobs in industrial cities or off-farm activities. Income diversification is the process by which farmers build a diversified source of income from activities aimed at improving the household's well-being, social support and risk management. Income generation is one of the components of a livelihood strategy. From time immemorial, rice has always been the main agricultural crop and the main product of Vietnamese agriculture. Rice occupies 60% of agricultural land and is the main cultivation direction of each producing household. Therefore, farmers immediately lose their primary income when the crop fails, and they fall into poverty. Recognizing the dangers of this problem, in recent years of renovation, with the guidance and support of the state, Vietnamese farmers have enormously changed production and diversified agricultural products. In each farmer's household, in addition to growing rice, they also grow crops (potatoes, corn, beans), industrial crops for seeds (coffee, pepper, cashew), perennial fruit trees and timber trees. In addition, livestock development includes raising fish, buffaloes, cows, goats, pigs and poultry. In addition, many new products have appeared, such as mushrooms, ornamental flowers and medicinal herbs. In recent years, many other agricultural products of Vietnam have risen to highly competitive rankings.

Among the above examples of coping with financial shortages, there is the emergence of bricolage due to the networks and social relations of farmers. For instance, formal networks of farms such as the production value chain, women's, farmer's or agricultural extension's unions, agricultural cooperatives, etc., were established in Vietnam a long time ago to implement politic missions and support farmers in cooperating in agriculture production, sharing information, knowledge and experiences, supporting farmer members with more access to markets,

2 See: https://thuvienphapluat.vn/van-ban/Quyen-dan-su/Bo-luat-dan-su-2015-296215.aspx.

technology, financial resources and material input sources, etc. Overtime, the participation of farmers in these networks, initially due to obligations to local authorities, helps them to have close relationships with other members of the network. These close relationships and, subsequently, trusting relations provide farmers with more access to all of the necessary inputs that they need, including financial resources collected in a formal as well as informal manner. The existence and rapid spread of "Hui", an informal way of mobilizing capital, is evidence for trusting relationships among farmers that can be established from the above networks or from other social relations of farmers. Such networks may comprise the farmers in a geographical area, including the farmers and their customers, suppliers, competitors who have farming activities in the same area. "Hui" can enable them to quickly obtain financial resources without collateral or many procedures and approval steps, such as bank loans, although this method of capital mobilization was illegal before 2015. This case confirms once again that two environmental factors, including the institutional environment and a community of friends with customers, suppliers and even competitors, are domains under which bricolage can take place, as argued by Baker and Nelson (2005) and Ronkko et al. (2014).

10.7. Human resources and skills

The agriculture sector is also seriously lacking workers. Young people in rural areas tend to work in industrial parks (Schnitze et al. 2016). Only older people still stick with farming work. They often carry out heavy work, such as harvesting or pulling the water hose throughout the field. In that case, to maintain critical farm activities, farmers mainly rely on their family members, relatives or friends in their social circles and seasonal labor as low-cost alternatives.

Farmers often have limited educational backgrounds, which prevents them from planning ahead and they thus have less ability to deal with shocks. A farm headed by a relatively well-educated person may be more likely to understand the benefits of new seed varieties and how to employ them to achieve improvement in yield rightly. The learning needs of farmers have significantly increased, most of which are self-learning through networking. Community-based informal training activities on indigenous agricultural practices can also be an essential aspect of the agricultural system of innovation. As a farm member of formal networks including agricultural cooperatives, women's, farmer's or agricultural extension unions and production chains at each province in Vietnam, farmers can obtain information, knowledge, experiences and skills shared by formal agriculture extension workers and other farm members. These organizations provide their farm members with consulting agricultural services in case there is a new agricultural policy, program and project.

They also follow the crop in each season and give regular advice or requirements in terms of seeds, agricultural care procedures, fertilization, harvesting and postharvest techniques, etc. In some cases, training courses will be organized by such organizations to update new knowledge and technology for farmers. Such formal networks thus become an important source for sharing and transferring knowledge.

Figure 10.6. *Farmers guided by instructors from the Vuong Thanh Cong company (source: https://daklak24h.com.vn/nong-nghiep/78795/lop-hoc-mien-phi-ve-san-xuat-ca-phe-huu-co.html). For a color version of this figure, see www.iste.co.uk/casadella/agri-innovations.zip*

The production chain also appears as another formal network that is responsible for improving farming skills and experiences for farmers. For example, the Vuong Thanh Cong Company in Dak Lak and Phus Sinh in Son La have opened dozens of free classes for hundreds of farmers to spread organic and sustainable coffee production processes to enhance the value of the coffee chain. According to the companies' CEO, farmers want to cultivate organically. However, most farmers still follow the inorganic direction because they can increase production. Therefore, it is necessary to help people fully understand the organic coffee production process to create green-quality products for consumers. At another formal scale related to production chains, Lien-Viet-Post Bank, as mentioned above, provides capital for farmer-borrowers, but also advises on plant varieties and planting techniques, and commits to purchasing by primary agents if the farmers strictly follow the recommended agricultural care procedures, creating a basis for farmers to feel secure in farming. This case strongly confirms the emergence of bricolage thanks to the environmental domain: customers. Furthermore, due to the prevalence of "word of mouth" culture, in an informal manner, the close relations of Vietnamese farmers with other members in their social circles with farmers also enable the easy sharing and transfer of knowledge and experiences about farming activities, regardless of whether they are good or bad.

10.8. Conclusion

Bricolage is considered the reconstruction and reuse of underutilized resources to produce new meaning, purpose and value (Salunke et al. 2013). Under scarcity, the novel combination of "nonstandard resources" can sometimes "reach brilliant unforeseen results" (Lévi-Strauss 1967, p. 17). Bricolage is exhibited in different ways to address resource scarcity in the agriculture sector. The dominant pattern is that farmers creatively and flexibly reallocate existing farm resources for new use or exploit all resources at hand, including land, income and labor, which are inexpensively available. Bricolage at any level (from individual to organization and interorganization) is characterized by some common actions: making do, refusing to enact limitations, recombining all resources available for new purposes to create something that is suitable for the needs of bricoleurs with little regard for regulations or standards. Bricolage may manifest in the following three main environmental domains: (1) inputs, consisting of physical inputs, labor and skills; (2) customers or markets; and (3) institutional and regulatory environment (Baker and Nelson 2005; Ronkko et al. 2014). The analysis of literature on bricolage in agriculture in some African countries has highlighted these actions of bricolage and environmental domains that stimulate bricolage. This chapter follows this framework to observe the emergence of bricolage in Vietnamese agriculture sectors. The patterns that the Vietnamese farmers follow with each resource constraint (including infrastructure and technology, financial resources, and human resources and skills) show that bricolage emerges in a wide range depending on each environmental domain. However, the role of bricolage in limiting environmental issues is not very strong, even though agriculture is Vietnam's second largest source of GHG emissions after manufacturing.

10.9. References

Anderson, J., Learch, C.E., Gardner, S.T. (2016). National survey and segmentation of smallholder households in Uganda: Understanding their demand for financial, agricultural, and digital solutions. CGAP Working Paper, Washington, DC.

Baker, T. (2007). Resources in play: Bricolage in the toy store(y). *J. Bus. Venturing*, 22(5), 694–711.

Baker, T. and Nelson, R.E. (2005). Creating something from nothing: Resource construction through entrepreneurial bricolage. *Administrative Science Quarterly*, 50(3), 329–366.

Benouniche, M., Zwarteveen, M., Kuper, M. (2014). Bricolage as innovation: Opening the black box of drip irrigation systems. *Irrigation and Drainage*, 63(5), 651–658.

Cunha, M.P., Rego, A., Oliveira, P., Rosado, P., Habib, N. (2014). Product innovation in resource-poor environments: Three research streams. *Journal of Product Innovation Management*, 31(2), 202–210.

Desa, G. (2012). Resource mobilization in international social entrepreneurship: Bricolage as a mechanism of institutional transformation. *Entrepreneurship Theory and Practice*, 36(4), 727–751.

Duong, P.B. and Izumida, Y. (2002). Rural development finance in Vietnam: A microeconometric analysis of household surveys. *World Development*, 30(2), 319–335.

Duymedjian, R. and Rüling, C.-C. (2010). Towards a foundation of bricolage in organization and management theory. *Organization Studies*, 31(2), 133–151.

Faure, G., Knierim, A., Koutsouris, A., Ndah, H.T., Audouin, S., Zarokosta, E., Wielinga, E., Triomphe, B., Mathé, S., Temple, L., Heanue, K. (2019). How to strengthen innovation support services in agriculture with regard to multi-stakeholder approaches. *Journal of Innovation Economics & Management*, 28(1), 145–169.

Garud, R. and Karnøe, P. (2003). Bricolage versus breakthrough: Distributed and embedded agency in technology entrepreneurship. *Research Policy*, 32(2), 277–300.

Ha, T.T.P., van Dijk, H., Bosma, R., Sinh, L.X. (2013). Livelihood capabilities and pathways of shrimp farmers in the Mekong Delta, Vietnam. *Aquaculture Economics & Management*, 17(1), 1–30.

Halme, M., Linderman, S., Linna, P. (2012). Innovation for inclusive business: Intrapreneurial bricolage in multinational corporations. *Journal of Management Studies*, 49(4), 743–784.

Kilelu, C., Klerkx, L., Leeuwis, C. (2017). Supporting smallholder commercialisation by enhancing integrated coordination in agrifood value chains: Experiences with dairy hubs in Kenya. *Experimental Agriculture*, 53(2), 269–287.

Levi-Strauss, C. (1967). *The Savage Mind*. University of Chicago Press.

Mahajan, S. (2013). Women entrepreneurship in India. *Global Journal of Management and Business Studies*, 3(10), 1142–1148.

Minh, T.T., Larsen, C.E.S., Neef, A. (2010). Challenges to institutionalizing participatory extension: The case of farmer livestock schools in Vietnam. *Journal of Agricultural Education and Extension*, 16(2), 179–194.

Musona, J., Sjögrén, H., Puumalainen, K., Syrjä, P. (2020). Bricolage in environmental entrepreneurship: How environmental innovators "make do" at the bottom of the pyramid. *Business Strategy & Development*, 3(4), 487–505.

Rönkkö, M., Peltonen, J., Arenius, P. (2014). Selective or parallel? Toward measuring the domains of entrepreneurial bricolage. In *Entrepreneurial Resourcefulness: Competing with Constraints*, Corbett, A.C. and Katz, J.A. (eds). Emerald Group Publishing Limited, Bingley.

Rosenzweig, S., Grinstein, A., Ofek, E. (2016). Social network utilization and the impact of academic research in marketing. *International Journal of Research in Marketing*, 33(4), 818–839.

Salunke, S., Weerawardena, J., McColl-Kennedy, J.R. (2013). Competing through service innovation: The role of bricolage and entrepreneurship in project-oriented firms. *Journal of Business Research*, 66(8), 1085–1097.

Sboui, O., Kefi, S., Dubois, M.J., Rizzo, D. (2022). Bricolage of agri-equipment. *International Conference of Technologies and Solutions for Efficient and Sustainable Agriculture*, November, Villepinte. 5th SIMA AgriTechDay.

Schnitze, H., van Thanh, T., Thao, N.T., Braunegg, G. (2016). An integrated eco-model of agriculture and small-scale industry in craft villages toward cleaner production and sustainable development in rural areas – A case study from Mekong delta of Viet Nam. *Journal of Cleaner Production*, 137, 274–282.

Senyard, J., Baker, T., Steffens, P., Davidsson, P. (2014). Bricolage as a path to innovativeness for resource-constrained new firms. *Journal of Product Innovation Management*, 31(2), 211–230.

Tarchiani, V., Rossi, F., Camacho, J., Stefanski, R., Mian, K., Pokperlaar, D., Coulibaly, H., Sitta Adamou, A. (2017). Smallholder farmers facing climate change in West Africa: Decision-making between innovation and tradition. *Journal of Innovation Economics & Management*, 24, 151–176.

Tindiwensi, C.K., Abaho, E., Munene, J.C., Muhwezi, M., Nkote, I.N. (2020). Entrepreneurial bricolage in smallholder commercial farming: A family business perspective. *Journal of Family Business Management*, 11(4), 423–439.

Tsoraeva, E., Bekmurzov, A., Kozyrev, S., Khoziev, A., Kozyrev, A. (2020). Environmental issues of agriculture as a consequence of the intensification of the development of agricultural industry. *E3S Web of Conferences*, 215. EDP Sciences. doi: 10.1051/e3sconf/202021502003.

Vilchez, V.F., and de la Hiz, D.I.L. (2018). Lessons on frugal eco-innovation: More with less in the European business context. In *The Critical State of Corporate Social Responsibility in Europe*, Tench, R., Jones, B., Sun, W. (eds). Emerald Publishing Limited, Bingley.

World Bank (2022). No time to waste: The challenges and opportunities of cleaner trade for Vietnam [Online]. Available at: https://openknowledge.worldbank.org/bitstream/handle/10986/36819/No-Time-to-Waste-The-Challenges-and-Opportunities-of-Cleaner-Trade-for-Vietnam.pdf?sequence=1&isAllowed=y.

Wu, L. and Liu, H. (2022). How bricolage influences green management in high-polluting manufacturing firms: The role of stakeholder engagement. *Business Strategy and the Environment*, 31(7), 3616–3634.

11

The Contribution of Food Hubs in the Digital Age to the Sustainable Agri-food Transition: A Review of Research in Sub-Saharan Africa

11.1. Introduction

According to the Food and Agriculture Organization of the United Nations (FAO) (2017), in the face of the challenges of the 21st century (urbanization, urban population growth, food and nutritional security), it is urgent to rethink the ways of organizing food supply. This urgency is noted in sub-Saharan African countries, where they are confronted with certain shocks such as the Covid-19 health crisis and the war in Ukraine, which bring to the forefront the fragility of the conventional agri-food model and its distribution circuit (Dury et al. 2021; Glauben et al. 2022). Thus, the food crisis generated by these shocks is not spontaneous but rather the result of a long struggle for the governance of the conventional agrifood system (Canfield et al. 2021). In the face of this, initiatives are developing around the governance of sustainable agri-food models (certified and non-certified organic agriculture, agroecology) as alternatives (Diop 2007; Bellon 2016; Bendjebbar 2018) to the conventional system. Their alternative characteristics are related to several elements, such as production techniques, product quality in relation to the territory and marketing circuit (proximity between the producer and the consumer). They contribute to the diversification of the food supply and boost the local or territorial market to which the industrial agri-food system alone seems to have difficulty responding (Temple et al. 2017).

Chapter written by Gérard DE LA PAIX BAYIHA.

Their development in sub-Saharan Africa is all the more relevant because they are part of a context marked by (i) agricultural sustainability issues; (ii) a galloping urban population and growing urbanization (Leridon 2015) that is expected to exert significant pressure on the Natural Environment, leading to resource conflicts (Suckwani et al. 2019); and (iii) national value chains that are characterized by the presence of several intermediaries (Abate et al. 2021). These three points affect the demand, distribution and consumption pattern of food products (Thornton et al. 2011; Ortega et al. 2017).

In addition, in sub-Saharan Africa, approximately 80% of small-scale farmers, characterized by plots of land averaging 2 ha, who are located in rural areas, contribute to nearly 90% of food production in some countries in the SSA region (Wiggings et al. 2013). As such, they play an important role in feeding people in urban areas. These small-scale farmers mobilize practices related to sustainable agri-food models by choice or by constraints (Bendjebbar 2018; Bayiha et al. 2019; Haggar et al. 2021). They market their production through traditional markets that they compose with small traders in a specific location (Davies et al. 2022). These markets are essential for supplying cities in sub-Saharan African countries. Despite the importance of these markets, they have difficulty marketing their products to cities (Abate et al. 2021; Haggar et al. 2021). This marketing problem is related to various factors. For example, we have geographical distance, technological and infrastructural problems (road, storage and processing units), information asymmetry on market prices and consumer food preferences, competition with supermarkets and retail stores, and poorly located and designed markets that do not allow adequate access to quality food for urban consumers (Janin 2019; Walther et al. 2019; Abate et al. 2021; Davies et al. 2022).

Due to all of the above, there is a major challenge of having innovative marketing models to sustainably supply cities in sub-Saharan African countries with products from sustainable agri-food models (Lutz and Schachinger 2013). This observation was also noted by the French Development Agency and the World Bank in 2020 (Lalineau et al. 2020). To this end, the concept of territorial innovation proposes possible solutions (Faure et al. 2018). In agriculture, it allows for the implementation of sustainable and local food governance modes, that is, modes of coordination of production and transportation until their distribution on a territorial scale of products from sustainable agri-food models (Gonçalves and Zeroual 2016). This innovation is based on the proximity between the producer and the consumer on a territorial scale to facilitate the distribution of products. We thus note the development of short food circuits defined as forms of sales mobilizing at most one intermediary between the producer and the consumer (Le Coq et al. 2019). The small number of intermediaries leaves the choice to the actors on different modes of distribution. For example, we can have the "baskets of goods", which concern the distribution of baskets of products, in the form of prepaid subscriptions, by groups of consumers who participate in the management of orders, distribution and even production (Mollard and Pecqueur 2007); the Associations for the Maintenance of a Peasant Agriculture (*Associations pour le Maintien d'une Agriculture Paysanne* (AMAPs)), which aim to base the emergence of the territorial quality rent, that is, the joint valuation of quality products and demand on a territory; and the "food hubs" that achieve the "aggregative scaling" of local quality and sustainable products different from the economies of scale of the conventional agri-food system in order to maintain their positive environmental, social and economic impact within a territory (Berti and Mulligan 2016). Some of these modes are found in cities in sub-Saharan African countries. We have, for example, the organic food baskets in Nairobi, Kenya (Tankam 2015), and the AMAPs for sustainable agriculture in the Provence area of Senegal (Gassama 2021). Despite this presence, these circuits are still struggling to expand in sub-Saharan Africa (Janin 2019) to contribute to an "aggregative scaling" of sustainable and quality products.

Thus, among a set of short food circuits in sub-Saharan Africa, this work will focus on the development of food hubs. The main reason for this choice is that, unlike AMAPs and food baskets, which focus primarily on the quality of the product in relation to the territory and the direct sale to consumers, the "food hubs" focus on the development of additional actions. They provide solutions to the challenge of scaling up sustainable agri-food systems with strong infrastructural and organizational needs of small-scale producers in sub-Saharan Africa to supply cities with high quantities of sustainable and quality products (Mollard and Pecqueur 2007; Berti and Mulligan 2016). Moreover, they are different from the wholesalers and buying groups which have existed for many years because the latter are key actors intervening in the food supply modes of long circuits (Gonçalves and Zeroual 2013). For this purpose, "food hubs" are points of entry to the market for small farmers from rural areas that are involved in sustainable agri-food models and have difficulties marketing their production to cities (Clark et al. 2019). Since their emergence in the United States in the 1970s, there has been no unanimous definition by different actors (researchers, policy makers) (Horst et al. 2011). The main reason is that they aim to meet the needs of a population rooted in a specific place, giving each "food hub" a unique expression (Clark et al. 2019). They are mostly defined as "businesses or organizations that actively manage the aggregation, distribution, and marketing of food products, with identified origins, from local and regional producers, to enhance their ability to meet wholesale, retail, and institutional demand" (Fischer et al. 2015).

Given the evolution of digital technology referring to information and communication technologies (ICT) (laptop, computer and their connectivity through the Internet) in Africa it can be a solution to explore to develop a new economic model of exchange for some African agri-food systems that are not well integrated (Fielke et al. 2020; Huet and Moriniere 2021). For example, in Benin, information sharing through digital technology contributes to the agroecological transition process (Paget et al. 2022); in Burkina-Faso, it contributes to the establishment of a marketing platform for agropastoral products (Nubukpo et al. 2022). Digital technology is thus a catalyst for the development of food hubs in the process of transitioning to sustainable agrifood systems in a territory (Sgroi and Marino 2022). To this end, we *hypothesize that food hubs in the era of digital expansion in sub-Saharan Africa offer an opportunity to market products from existing sustainable agri-food models within a territory.*

From this perspective, this work proposes to analyze how different knowledge communities (Conein 2004) mobilize the concept of the "food hub" in the context of digital expansion for the development of sustainable agri-food models in sub-Saharan Africa.

The objective is to review the literature on food hubs in sub-Saharan Africa to highlight the realities and challenges of their emergence in the process of promoting sustainable agri-food models.

This chapter is organized into three parts. The first part presents the literature review on territorial innovation and the contribution of food hubs to the transformation of agriculture and agri-food. The second part highlights the methodology and the main results. The third part discusses these results.

11.2. Literature review

This section is subdivided into two sections. The first highlights the importance of territorial innovation for the development of short food circuits. The second presents an innovative short circuit, namely, the "food hubs".

11.2.1. *Territorial innovation, a lever for the development of short food circuits*

The notion of territorial innovation seeks to highlight the relationship between innovation and territory. In this notion, the concept of territory, generally on the local or regional scale of a country, is defined as a space of governance mobilizing at the same time the geographical (kilometer distance between two elements), socioeconomic, organizational and institutional proximity of individuals and cities (Loilier 2010; Billaudot 2022). This proximity facilitates, on the one hand, the sharing of information and creates a climate of trust between actors (Loilier 2010) and, on the other hand, the development of healthier and more sustainable food

(Holloway et al. 2007). Territorial innovation can be broken down into two aspects: innovative territories and the territorialization of innovation (Soulard et al. 2018). We will mobilize it in this work according to the second aspect. It is materialized by the implementation of new modes of governance and new political arrangements at the local or territorial scale (Rey-Valette et al. 2014). The main reason for this choice is due to the fact that, because several actors intervene at different levels of the food supply, from production to consumption (Lutz and Schachinger 2013), one of the main challenges of the transition to a sustainable "agri-food" system is having new modes of arrangements and governance at the territorial level. The "agri-food" system is defined as a sociotechnical system[1] bringing together different actors at the territorial level: farmers, research, upstream and downstream actors, public policies, consumers and civil society (Lamine 2012). These actors can be summarized into three types: civil society, market actors (small farmers, industrialists, distributors) and the state (public authorities and/or local communities). Thus, in the context of the transition to a sustainable agri-food system, food governance, defined as an approach to coordinating actors with the aim of promoting their organization and reducing the atomization of initiatives affecting food systems (Billaudot 2002), is relevant. In the transition process, the definition of new modes of food governance by the state can be seen at two levels. Firstly, it will facilitate transactions of local sustainable agri-food products based on the sharing of information directly or through an intermediary (processor, transporter, distributor via a system or point of sale) between small farmers in rural areas and consumers in urban areas. The information will include, for example, pricing systems, product quality and constraints influencing production and consumption. Secondly, the state, through a planning process, can promote the development of short food circuits to improve access to sustainable agri-food products for consumers (Bouroullec 2020). These circuits have several advantages on the economic, social and environmental levels within a territory (Moustier 2017). For example, on the economic level, they contribute to a better remuneration of producers due to the reduction of intermediaries and to a better price formation mechanism thanks to the sharing of information on product quality. On the social level, they contribute to the creation of a larger community of producers and consumers of quality products. Environmentally, they reduce pollution. In other words, they contribute to the organization of sustainable agri-food chains and markets. Thus, among different models of short food circuits (baskets of goods, AMAP, "food hubs"), we focus on "food hubs".

1 A *sociotechnical system* is defined as the interaction between several elements (actors, market, technology and institutions) that fulfill a societal need (Geels 2004). In this study, this need refers to the valorization of products from sustainable agri-food models from rural areas to cities in sub-Saharan Africa via a short food circuit. In addition, technology refers to digital technology.

11.2.2. *Food hubs, an innovative model for supplying cities in the transition process toward sustainable agri-food systems*

In the context of the transition to sustainable agri-food systems, the governance of the supply chain of sustainable agri-food products from rural areas by the state in cities, to be effective, requires innovative models. Thus, models such as food hubs are traditionally considered in the literature as levers to build socially just and more economically and ecologically robust agri-food systems that connect small farmers to consumers as much as possible. There is a typology that represents a wide range of food hub models. This typology can be differentiated on the basis of several criteria: stakeholders, function, purpose, type of products, how the function and operation of food hubs influence other actors in the agri-food system, different levels of farmer control over their market relationships, scale and the legal structure through which food hubs operate in areas such as capital investment, risk management and liability exposure (Morley et al. 2008). On this basis, the work of Horst et al. (2011) identified nine types of food hubs. This typology can be summarized into three main groups (Matson et al. 2013; Sgroi and Marino 2022). The first is the direct-to-consumer model, which represents food hubs through which consumers pay in advance for products from producers. The second is the plantation (agricultural cooperative) to enterprise model, which characterizes food hubs that connect producers to large-scale intermediaries such as restaurants and retailers rather than directly to consumers. The third is the hybrid model that combines the two previous models. Food hubs thus help provide an economically viable market for small farmers growing these types of products (Day-Farnswortha and Morales 2011; Berti and Mulligan 2016). As such, they are a model for supporting sustainable "agri-food" systems (Cleveland et al. 2014).

Therefore, with the advancement and lowering of costs related to ICT on a global scale, the consideration of "food hubs" as only physical marketing centers to evolve to another form based on digital distribution. Thus, there is the birth of digital platforms that facilitate transactions for obtaining products by the consumer, making them easier to set up as well as at low prices (Berti and Mulligan 2016). "Digital food hubs" are then defined as "Internet-based online directories and marketplaces that promote effective connections between producers and consumers of local and regional foods, including institutions, restaurants, and stores. They may have a physical location, but this is not necessary" (Horst et al. 2011). They participate in the transition to sustainable "agri-food" systems (Sgroi and Marino 2022). There are two types of digital food hubs. The first represents virtual spaces through which producers, consumers and even processors network to interact to engage in commercial exchanges. These exchanges do not take place in these virtual spaces. Rather, they are directly between consumers and sellers. The second type is materials that contribute to information sharing between consumers and sellers on an ongoing basis. They give the possibility to make online orders to process payment and to coordinate delivery mechanisms. This type is more complex to implement.

Digital food hubs, mainly driven by public actors and local governments, are more common in developed countries such as the United States, with an emergence in some European countries (Sgroi and Marino 2022). They are an essential component in the digital era for the scaling up of local food systems (Fardkhales and Lincoln 2021) and thus a transition toward "sustainable agri-food" systems (Sgroi and Marino 2022). Faced with the challenges of "food hubs", in 2020, the European Union financed the implementation of the project "FOODLAND"[2] in Africa. It aims to improve the diversity of food production and consumption for a nutritional transition in six African countries, four of which are located in sub-Saharan Africa (Tanzania, Kenya, Ethiopia, Uganda). By 2022, the project has enabled the establishment of 10 "food hubs" in the cities of these countries. FOODLAND defines a "food hub" as "a community of local operators (actors and organizations), that is, small-scale farmers, agribusinesses, the state, researchers and non-governmental organizations, aiming to make joint decisions on research and development and to enable the adoption of innovations". This project operates with two main approaches. The first refers to a "bottom-up" approach. This focuses on analyzing the needs and choices of small-scale farmers and consumers from a gender perspective. The second focuses on participatory methods and training courses for small farmers and food processors to improve their activities.

Given these dynamics, it is therefore interesting to review the literature on food hubs in sub-Saharan Africa. This review will allow us to see to what extent they are mobilized by knowledge communities to overcome the difficulties that disrupt small rural farmers from reaching the growing demand for quality products within the cities of sub-Saharan African countries (Tankam 2015).

11.3. Methodology of the documentary research and results

Here, we question the state of the art of the works on the contribution of "food hubs" and digital hubs in the development of sustainable agroalimentary models in sub-Saharan Africa. It is based on a literature search. We chose not to establish it through the most widely used human and social science research databases, such as Web of Science and Scopus, because we do not have access to these databases, which are not free. We therefore focused on an alternative database to these: Google Scholar. It has the advantage of being free and covers a large variety of scientific works from knowledge communities: peer-reviewed and nonpeer-reviewed articles, theses, reports and books (Kosmopoulos et al. 2008). Even if it has some limitations

2 Available at: https://foodland-africa.eu/foodhubs/.

(uncontrolled content of works and citations, proliferation of publication sources, several versions of the same article), it seems interesting to exploit it, but with a rigorous methodology. In light of this, we convened the five-step approach of Fahimnia et al. (2015).

The first was to choose the appropriate search terms for the research. The keywords used exclusively in English are "food hubs", "food hub" and "Africa". A combination of these words led to the definition of two search equations. The first is "food hubs" OR "food hub" AND Africa*, and the second is formulated as follows: "food hubs" OR "food hub". We have chosen not to use the French translation of "food hubs", which is not yet stable in the French literature. In addition, it is a terminology that is used in the French literature.

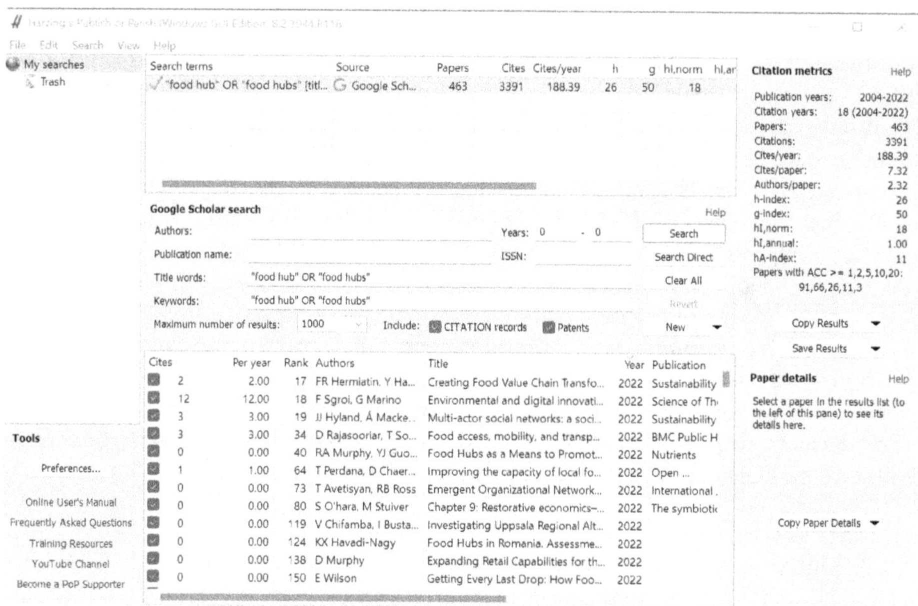

Figure 11.1. *Publications of studies related to "food hubs" or "food hubs" on Google Scholar. For a color version of this figure, see www.iste.co.uk/casadella/agri-innovations.zip*

The second step was to perform this title and keyword search in Google Scholar. This was done on September 30, 2022. For this purpose, the "Publish or Perish" (PoP) version 8.2.3944 for Windows was mobilized to compile the results of the studies (journal and conference articles, reports, theses) obtained. For the first equation, no studies were obtained. The second search equation produced 463 studies. For each study, the software provides the number of citations, authors' names, year and place of publication. The published works were found to cover the period from 2004 to 2022 (Figure 11.1).

The third step was to refine the search. It was done manually and automatically. Before doing so, we had to define the interval of years for the data collection. Given the small number of studies obtained, we kept the initial interval proposed by the PoP software. These results were exported in RIS format to include the relevant information (names and affiliations of the authors, abstract, keywords and references). Afterwards, we imported these studies into Zotero software. The latter allowed us to (1) automatically highlight only peer-reviewed journal articles (conference papers, books and book chapters being excluded); (2) manually merge articles that were duplicates or more; and (3) check the articles that were trustworthy. This work provided us with 105 articles spanning the period 2011–2022. We manually checked in each article, the abstract and the title the studies that dealt with the "food hubs" or the "food hub" in sub-Saharan African countries, on the one hand, and that highlighted the digital ones, on the other hand. It was found that none exist. It took place from September 30 to October 3, 2022. Notwithstanding this dynamic of publications on this innovative model (Figure 11.2), no articles targeting sub-Saharan African countries were identified.

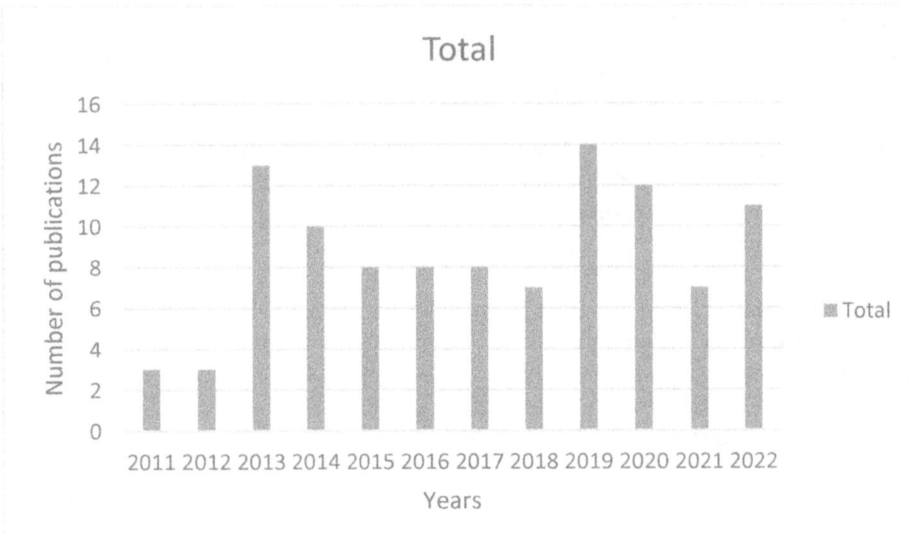

Figure 11.2. *Change in the number of publications referring to the concepts "food hubs" and "food hubs" between 2011 and 2022 (N = 105). For a color version of this figure, see www.iste.co.uk/casadella/agri-innovations.zip*

The objective would have been to build an analysis grid around seven pieces of information: when was the food hub implemented? Why was it set up and how is it perceived by the state? Are there any digital food hubs? What role does digital technology play in its development? What are the brakes and levers for its implementation? What are the techno-economic-political knots that make the "food hub" currently subject to discussion? What is its contribution to the supply of sustainable agri-food products to cities? The last two steps (compilation of preliminary statistics and data analysis) were not convened.

11.4. A lack of scientific work on food hubs in sub-Saharan Africa despite their development

Food hubs are being developed with the aim of promoting sustainable "agri-food" systems (Berti and Mulligan 2016). Despite this, to date, no research articles have been conducted on them in sub-Saharan Africa. However, since January 2022, there has been the establishment of 10 "food hubs" in cities in four countries located in sub-Saharan Africa (Ethiopia, Uganda, Kenya, Tanzania) to promote the development of sustainable agriculture within the territories where they are located. Since the beginning of the project that led to their establishment, it has been noted that the public authorities are not yet involved. As a result, there are still no local institutional mechanisms to support their development. However, the state plays an important role in the relationship between innovative projects and their development so that the results benefit the whole community beyond direct actors (Casadella and Temple 2016). However, steps are being taken. Each of these food hubs does not specifically focus on the aggregation and distribution of food products, which are the main functions of food hubs. They have additional actions not emphasized in the types of food hubs (direct-to-consumer model, farm-to-business cooperative model, hybrid model, digital food hub) identified in the literature (Horst et al. 2011; Matson et al. 2013; Sgroi and Marino 2022). These actions are mainly oriented toward the development of technological and process innovations that will impact sustainable agri-food systems at three levels. Firstly, production, through technologies that will contribute to increasing yields in a sustainable manner, will lead to the aggregation of farmers who will contribute to increasing the food supply. Then, the transformation occurs through the implementation of processes that will contribute to the commitment of food processors to local food hubs. Finally, we have the consumer through the implementation of tools that will encourage them to buy more and better quality products. For example, the Laelay Machem food hub in Mekele, Ethiopia, is interested in technological innovations in precision harvesting, processing and packaging of composite flours (teff and moringa) and oil (moringa) to ensure quality. In the same country, the one located in Addis Ababa, which focuses on fresh vegetables and legumes, pays attention to two levels within these chains. The first is at the production level, promoting technological innovations that will improve sustainability and productivity. The second level is to facilitate their commercialization on the markets. However, the "food hub" located in Kamuli, Uganda, aims to promote technological innovations that will contribute to the agroecological intensification of important crops (maize, beans, coffee, soybeans, sweet potatoes) in this territory. These innovations (agroecological practices, milling, storage) will contribute to solving different problems (sanitary diseases, high prices of chemical inputs, reduction of incomes) encountered by the stakeholders within these sectors. In light of this, food hubs in sub-Saharan Africa seem to operate at two levels. The first is to promote technological innovations to respond to a particular problem in a locality at the level of production and packaging with the aim of having a good quality and sustainable agricultural product. The second is to eventually contribute to their aggregation and marketing. Given the new character of these food hubs and the lack of information on them in Africa, it is still difficult at this stage to classify them according to the three main types identified in the literature.

Although in the contexts (developed countries) where the "food-hubs" have been installed for a few years, they are more oriented toward the aggregation, the marketing of the products through mainly the model of direct sale to the consumers, the model of the agricultural cooperative to the company, the digital "food hub". For example, regarding the first model, there is the "Puget Consumers Co-op" located in Washington in the United States. It develops retail outlets throughout the region through which members and non-members can shop for local organic food products (Horst et al. 2011). In Europe, we find situations in which digital technology is mobilized to strengthen the development of food hubs. For example, we have the case of a food hub located in Parma, Italy. It mobilizes digitalization to mitigate the negative impacts of the distribution circuit of the conventional system within the city (Morganti et al. 2015). The expansion of digitalization helps to fill gaps in the local food infrastructure, help consumers find local agri-food products and support new forms of food retailing. These different operationalizations of food hubs in sub-Saharan Africa and in developed countries confirm the fact that each one gives rise to a unique expression to address local priorities (Clark et al. 2019). As mentioned above, we have intuitively presented a comparative picture of food hubs in Africa and in developed countries. These are synthesized from six criteria (stakeholders, product type, scale, functions and operation, legal structure, purpose) of food hubs highlighted in the Table 11.1 below. The building blocks for fulfilling these criteria are drawn from two sources. For sub-Saharan Africa, due to their development in 2022, they are taken from the FOODLAND project website, and for developed countries, they are mainly taken from the synthetic work of Horst et al. (2011), Berti and Mulligan (2016) and Hermiatin et al. (2022). We added two criteria: the existence of local institutional arrangements and the use of digital technology. In total, we have eight criteria.

Criteria	Sub-Saharan Africa	Developed countries
Stakeholders	Small-scale farmers, food processors, government, researchers, consumers and non-governmental organizations	Small farmers, food processors, distributors, retailers, government, consumers
Type of products	Sustainable agri-food	Sustainable agri-food
Scale	Territory	Territory
Functions and Operation	1) Production (services to small farmers), processing through the development of storage, packaging and labeling systems 2) Aggregation, coordination and marketing 3) Mobilization of a "bottom-up" approach that focuses on the analysis of the needs of small farmers and consumers 4) Implementation of participatory methods and training courses for small farmers and food processors to improve their activities	1) Aggregation, coordination and marketing 2) Services: (i) product services to add value and provide facilities (storage); (ii) producer advisory services to support the transition to agro-ecological or organic practices and assist stakeholders in developing a "network of practices" 3) Practice web to provide learning and innovation services to all stakeholders involved in the agro-ecological or organic "practice network"
Local institutional arrangements	Not yet operational	Operational
Use of digital technology	Operational at the production level	Operational at the production and supply level
Legal structure	Innovation Center	1) Non-profit or for-profit company or organization 2) Cooperative 3) Private company
Purpose	1) Improve sustainable, quality food production and consumption 2) Create new market opportunities for small producers	1) Improve sustainable, quality food production and consumption 2) Create new market opportunities for small producers

Table 11.1. *Summary of the comparison of food hubs in Africa and in developed countries*

Table 11.1 highlights the difference between food hubs in sub-Saharan Africa and in developed countries according to eight criteria. It notes certain similarities, such as the stakeholders, the scale and the purpose. However, we also find some differences, such as local institutional arrangements, certain functions and operations, legal status and the use of digital technology. For example, concerning the last criterion, in sub-Saharan Africa, it is used mainly in "food hubs" to solve problems at the production level. We find that digital technology is not yet fully in evidence in food supply in food hubs. One reason for this may be the fact that this innovative model was implemented on the continent in 2022. Thus, regarding the application of digital technology within food hubs on production, we can take the example of the one located in the Nakaseke district in Uganda. It offers digital solutions to small farmers to enable them to program their irrigation system to avoid wasting water. In developed countries, digital technology is contributing to food hubs, especially in food supply. For example, in Europe, we have food hubs located in the cities of Stroudco and Dean Forest in England, and these allow easy and affordable access of local food to consumers (Quarantia and Salvia 2017). The differences that exist between "food hubs" are explained in the work of Horst et al. (2011). They highlight that food hubs are used in different ways depending on the issues in a community to support local food systems.

11.5. Conclusion: a need for studies on food hubs in sub-Saharan Africa

In the processes of transition to sustainable agri-food systems in cities of sub-Saharan African countries, food supply is an important element. This is because urban consumers are increasingly concerned about food quality and safety (Ortega et al. 2017), and cities play an important role in the development of sustainable agri-food systems (Morgan and Sonnino 2010). For this, it is necessary to rethink the mode of food governance within territories to value short food circuits such as "food hubs" that serve to localize sustainable agri-food models. Thus, the "food hubs", whose presence is multiplying in developed countries, are an innovative model presented in the literature as a potential lever for this transition to take place. They are models for the supply of sustainable agri-food products on a territorial scale as an alternative to those proposed by conventional agriculture, which are subject to numerous criticisms. In the African context, little is known about how small-scale rural farmers distribute sustainable agri-food products to cities to address various challenges (market availability, physical and economic access, changing food and nutrition insecurity due to population growth, information asymmetry on product quality) faced by urban populations (Janin 2019; Lalineau et al. 2020). Food hubs will help address these challenges highlighted in the literature and by international organizations such as the French Development Agency (Agence Française de Développement) and the World Bank (Lalineau et al. 2020). For example, the adoption of innovative technologies to supply urban consumers with sustainable agri-food products will contribute to the improvement of their food and nutrition security (Adenle et al. 2019). With the expansion of digital technology in sub-Saharan Africa, food hubs offer a real platform to bring together urban consumers and small-scale farmers in rural or peri-urban areas and overcome the many challenges (storage, informational asymmetry on product quality for the consumer) that limit the marketing of sustainable agrifood products. In addition, information and communication technologies will allow this innovative model to make trade between these small producers and consumers in easier, faster and less expensive ways (Berti and Mulligan 2016). This work therefore aims to take stock of the scientific work done by different knowledge communities on this model in sub-Saharan Africa. We hypothesize that *food hubs in the era of digital expansion in sub-Saharan Africa offer an opportunity to commercialize products from existing sustainable agri-food models within a territory*. Despite the dynamic evolution of publications on food hubs in the literature since 2011, the literature search conducted reveals that there are no research articles on them published by different knowledge communities toward the African continent in general and the sub-Saharan African area in particular. This result can be explained by the recent emergence (year 2022) of "food hubs" on the continent. Their particularity is that to date, apart from the marketing function that is not yet visible due to their niche status, they focus mainly on technological innovations that will contribute to obtaining quality products. Thus, taking into account the fact that the AMAP and the baskets of goods that focus primarily on direct sales to consumers have difficulty developing in sub-Saharan Africa, the ideal type of "food hub" that could be encouraged is the model of the agricultural cooperative to the company. In this type, digital technology will play a key role as it evolves in Africa. This pioneering and exploratory work on this topic on the continent has confirmed this hypothesis.

Accordingly, this study suggests that case studies be conducted for each food hub that is already established within the territories of countries in sub-Saharan Africa. These studies can help inform stakeholders, including policymakers, on the critical factors for their success (Hylan et al. 2022). Thus, the objectives of this work will be twofold: (i) to analyze and describe the innovation processes of the "food hub" (agricultural cooperative to enterprise model) in terms of paradigm and technological trajectory. It will be a question of defining the relevant problems of the territory in terms of supplying cities with sustainable agri-food products and the specific knowledge that this type of "food hub" can contribute to their solution, and (ii) to conduct a cost–benefit analysis of their implementation.

11.6. References

Abate, G.T., Bernard, T., Janvry, A. (2021). Introducing quality certification in staple food markets in Sub-Saharan Africa: Four conditions for successful implementation. *Food Policy*, 105, 102173.

Adenle, A., Wedig, K., Azadi, H. (2019). Sustainable agriculture and food security in Africa: The role of innovative technologies and international organizations. *Technology in Society*, 58(August), 101143.

Balineau, G., Bauer, A., Kessler, M., Madariaga, N. (2020). Les systèmes agroalimentaires en Afrique. Repenser le rôle des marchés. Report, Agence française de développement et de la Banque mondiale, Paris/Washington DC.

Bayiha, G., Temple, L., Mathe, S., Nesme, T. (2019). Typologie et perspective d'évolution de l'agriculture biologique au Cameroun. *Cahiers agricultures*, 28, 3.

Bellon, S. (2016). Contributions croisées de l'agriculture biologique à la transition agroécologique. *Innovations agronomiques*, 51, 121–138.

Bendjebbar, P. (2018). Vers un modèle bio africain ? Trajectoires comparées d'institutionnalisation de l'agriculture biologique au Bénin et en Ouganda. PhD Thesis, Université Paris-Est.

Berti, G. and Mulligan, C. (2016). Competitiveness of small farms and innovative food supply chains: The role of food hubs in creating sustainable regional and local food systems. *Sustainability*, 8(7), 616.

Billaudot, B. (2002). Patrimoines productifs, secteur et territoire. *Géographie économie société*, 4(3), 259–303.

Bouroullec, M.D. (2020). Les différentes gouvernances des circuits courts alimentaires de vente en ligne collective. *Économie rurale. Agricultures, alimentations, territoires*, 371 (March), 59–75.

Canfield, M., Anderson, M., McMichael, P. (2021). UN food systems summit 2021: Dismantling democracy and resetting corporate control of food systems. *Frontiers in Sustainable Food Systems*, 5, 1–15.

Casadella, V. and Temple, L. (2016). Politiques d'innovation pour le développement des économies du Sud : regards croisés Sénégal/Cameroun [Online]. Available at: http://agritrop.cirad.fr/582715/.

Clark, J.K. and Inwood, S.M. (2016). Scaling-up regional fruit and vegetable distribution: Potential for adaptive change in the food system. *Agriculture and Human Values*, 33(3), 503–519.

Clark, J.K., Rouse, C., Sehgal, A.-R., Bailey, M., Bell, B.-A., Pike, S.-N., Sharpe, P.-A., Freedman, D.-A. (2019). A food hub to address healthy food access gaps: Residents' preferences. *Journal of Agriculture, Food Systems, and Community Development*, 9(1), 59–68.

Cleveland, D.-A., Muller, N.-M., Tranovich, A.-C., Mazaroli, D.-N., Hinson, K. (2014). Local food hubs for alternative food systems: A case study from Santa Barbara county, California. *Journal of Rural Studies*, 35(July), 26–36.

Conein, B. (2004). Communautés épistémiques et réseaux cognitifs : coopération et cognition distribuée. *Revue d'économie politique*, 113, 141–159.

Davies, J., Blekking, J., Hannah, C., Zimmer, A., Joshi, N., Anderson, P., Chilenga, A., Evans, T. (2022). Governance of traditional markets and rural-urban food systems in sub-Saharan Africa. *Habitat international*, 127(September), 102620.

Day-Farnswortha, L. and Morales, A. (2011). Satiating the demand: Planning for alternative models of regional food distribution. *Journal of Agriculture, Food Systems, and Community Development*, 2(1), 227–247.

Diop, A.-M. (2007). How can organic farming and agro-ecology contribute? *Africa Can Feed Itself*, 229–236.

Duru, M., Hazard, L., Magrini, M.-B. (2016). La santé comme cadre d'analyse pour penser conjointement les questions agricoles, environnementales et alimentaires. École-Chercheurs PSDR 4, Toulouse.

Dury, S., Alpha, A., Zakhia-Rozis, N., Giordano, T. (2021). Les systèmes alimentaires aux défis de la crise de la Covid-19 en Afrique : enseignements et incertitudes. *Cahiers agricultures*, 30, 12.

Fahimnia, B., Sarkis, J., Davarzani, H. (2015). Green supply chain management: A review and bibliometric analysis. *International Journal of Production Economics*, 162, 101–114.

Fardkhales, S.-A. and Lincoln, N. (2021). Food hubs play an essential role in the COVID-19 response in Hawaii. *Journal of Agriculture, Food Systems, and Community Development*, 10(2), 53–70.

Faure, G., Chiffoleau, Y., Goulet, F., Temple, L., Touzard, J.-M. (2018). *Innovation et développement dans les systèmes agricoles et alimentaires*. Éditions Quæ, Versailles.

Fielke, S., Taylor, B., Jakku, E. (2020). Digitalisation of agricultural knowledge and advice networks: A state-of-the-art review. *Agricultural Systems*, 180(April), 102763.

Fischer, M., Pirog, R., Hamm, M.-W. (2015). Food hubs: Definitions, expectations, and realities. *Journal of Hunger & Environmental Nutrition*, 10(1), 92–99.

Food and Agriculture Organization of the United Nations (2017). *The Future of Food and Agriculture: Trends and Challenges*. FAO, Rome.

Gassama, E.-B. (2021). L'AMAP en Provence : un modèle pour le développement des circuits courts alimentaires alternatifs au Sénégal ? PhD Thesis, Aix-Marseille Université.

Geels, F.-W. (2004). From sectoral systems of innovation to socio-technical systems. *Research Policy*, 33(6–7), 897–920.

Glauben, T., Svanidze, M., Götz, L., Prehn, S., Jaghdani, T.-J., Duric, I., Kuhn, L. (2022). The war in Ukraine, agricultural trade and risks to global food security. *Intereconomics*, 57(3), 157–163.

Gonçalves, A. (2013). La performance des organisations logistiques des circuits courts de distribution : une analyse des déterminants et leviers d'amélioration. PhD Thesis, Université de Lille 1.

Gonçalves, A. and Zeroual, T. (2016). Analyser les impacts des circuits courts alimentaires : une étude en Nord-Pas-de-Calais. *Revue de la régulation. Capitalisme, institutions, pouvoirs*, 20(December), 1–19.

Haggar, J., Nelson, V., Lamboll, R., Rodenburg, J. (2021). Understanding and informing decisions on sustainable agricultural intensification in Sub-Saharan Africa. *International Journal of Agricultural Sustainability*, 19(5–6), 349–358.

Harzing, A.-W.-K. and Van Der Wal, R. (2008). Google Scholar as a new source for citation analysis. *Ethics in Science and Environmental Politics*, 8(1), 61–73.

Hermiatin, F.-R., Handayati, Y., Perdana, T., Wardhana, D. (2022). Creating food value chain transformations through regional food hubs: A review article. *Sustainability*, 14, 8186.

Holloway, L., Kneafsey, M., Cox, R., Venn, L., Dowler, E., Tuomainen, H. (2007). Beyond the "alternative"–"conventional" divide?: Thinking differently about food production-consumption relationships. In *Alternative Food Geographies: Representation and Practice*, Maye, D., Holloway, L., Kneafsey, M. (eds). Elsevier, Oxford.

Horst, M., Ringstrom, E., Tyman, S., Ward, M.-K., Werner, V., Born, B. (2011). Toward a more expansive understanding of food hubs. *Journal of Agriculture, Food Systems, and Community Development*, 2(1), 209–225.

Huet, J.-M. and Moriniere, L. (2021). Le numérique pour sauver l'agriculture africaine ? *Futuribles*, 440(1), 65–80.

Hyland, J. and Hacken-Walsh, A. (2022). Multi-actor social networks: A social practice approach to understanding food hubs. *Sustainability*, 14(3), 1894.

Janin, P. (2019). Les défis de l'approvisionnement alimentaire : acteurs, lieux et liens. *Revue internationale des études du développement*, 237(1), 7–34.

Kosmopoulos, C. and Pumain, D. (2008). Révolution numérique et évaluation bibliométrique dans les sciences humaines et sociales. *Revue européenne des sciences sociales*, XLVI–141, 73–86.

Lamine, C. (2012). Changer de système : une analyse des transitions vers l'agriculture biologique à l'échelle des systèmes agri-alimentaires territoriaux. *Terrains & Travaux*, 1, 139–156.

Le Coq, J.-F., Sabourin, E., Bonin, M., Gresh, S.-F., Marzin, J., Niederle, P., Patrouilleau, M.-M., Vasquez, L. (2019). Politiques publiques d'appui à l'agro-écologie en Amérique latine : leçons et perspectives. In *La transition agro-écologique*, Cote, F.-X., Poirier-Magona, E., Perret, S. (eds). Éditions Quæ, Paris.

Leridon, H. (2015). Afrique subsaharienne : une transition démographique explosive. *Futuribles*, 407, 16.

Loilier, T. (2010). Innovation et territoire. Le rôle de la proximité géographique ne doit pas être surestimé. *Revue française de gestion*, 200(1), 15–35.

Lutz, J. and Schachinger, J. (2013). Do local food networks foster socio-ecological transitions towards food sovereignty? Learning from real place experiences. *Sustainability*, 5(11), 4778–4796.

Matson, J., Sullins, M., Cook, C. (2013). The role of food hubs in local food marketing (Rural Development Service Report 73). Report, U.S. Department of Agriculture, Washington DC.

Mollard, A. and Pecqueur, B. (2007). De l'hypothèse au modèle du panier de biens et de services. Histoire succincte d'une recherche. *Économie rurale. Agricultures, alimentations, territoires*, 300(September), 110–114.

Morgan, K. and Sonnino, R. (2010). The urban foodscape: World cities and the new food equation. *Cambridge Journal of Regions, Economy and Society*, 3(2), 209–224.

Morganti, E. and Gonzalez-Feliu, J. (2015). City logistics for perishable products. The case of the Parma's Food Hub. *Case Studies on Transport Policy*, 3, 120–128.

Morley, A.-S., Morgan, S.-L., Morgan, K.-J. (2008). Food hubs: The "missing middle" of the local food infrastructure? BRASS (ESRC) Research Centre, Cardiff University.

Moustier, P. (2017). Short urban food chains in developing countries: Signs of the past or of the future? *Natures Sciences Sociétés*, 25(1), 7–20.

Murphy, R.-A., Jacqueline Guo, Y., Cordeiro, H., Stroshein, S., Hamilton, C., Kozicky, S. (2022). Food hubs as a means to promote food security in post-secondary institutions: A scoping review. *Nutrients*, 14(19), 3951.

Nubukpo, K., Temple, L., Alexandre, C. (2020). Innovation numérique et transformation structurelle des économies africaines francophones, opportunités risquées pour le développement. *Technologie et innovation*, 5(3), 1–12.

Ortega, D.-L. and Tschirley, D.-L. (2017). Demand for food safety in emerging and developing countries: A research agenda for Asia and Sub-Saharan Africa. *Journal of Agribusiness in Developing and Emerging Economies*, 7(1), 21–34.

Paget, N., Nacambo, I., Fournier, S., Moumouni-Moussa, I. (2022). Traque des innovations numériques au service de la transition agroécologique au Bénin. *Cahiers agricultures*, 31, 1–10.

Palacios-Argüello, L., Sanchez-Diaz, I., Gonzalez-Feliu, J., Gondran, N. (2020). The role of food hubs in enabling local sourcing for school canteens. In *Food Supply Chains in Cities: Modern Tools for Circularity and Sustainability*, Aktas, E. and Bourlakis, M. (eds). Springer International Publishing, Cham.

Plumecocq, G., Debril, T., Duru, M., Sarthou, J.-P., Therond, O. (2018). Caractérisation socio-économique des formes d'agriculture durable. *Économie rurale*, 363(March), 99–120.

Prevost, B. (2012). Short food supply chains as social innovations for sustainable food in territories. *Norois*, 224, 7–20.

Quaranta, G. and Salvia, R. (2017). Participatory approach to a place-based sustainable rural development: E-market platform for a resilient agri-food chain. *Journal of Environmental Protection and Ecology*, 18(2), 616–622.

Rastoin, J.-L. (2018). Accélérer la transition vers une alimentation durable par un changement de paradigme scientifique et économique et des politiques publiques innovantes. *Systèmes alimentaires*, 3, 17–22.

Rey-Valette, H., Chia, E., Mathe, S., Michel, L., Nougarèdes, B., Soulard, C.-T., Maurel, P., Jarrige, F., Barbe, E., Guiheneuf, P.-Y. (2014). Comment analyser la gouvernance territoriale ? Mise à l'épreuve d'une grille de lecture. *Géographie, économie, société*, 16(1), 65–89.

Richard, F., Chevallier, M., Dellier, J., Lagarde, V. (2014). Circuits courts agroalimentaires de proximité en Limousin : performance économique et processus de gentrification rurale. *Norois, environnement, aménagement, société*, 230(June), 21–39.

Sgroi, F. and Marino, G. (2022). Environmental and digital innovation in food: The role of digital food hubs in the creation of sustainable local agri-food systems. *Science of the Total Environment*, 810, 152257.

Soulard, C.-T., Perrin, C., Scheromm, P., Chia, E., Clément, C., Michel, L., Hasnanoui Amri, N., Duffaut-Prévost, M.-L., Ubilla-Bravo, G. (2018). Les relations entre ville et agriculture au prisme de l'innovation territoriale. In *Innovation et développement dans les systèmes agricoles et alimentaires*, Faure, G., Chiffoleau, Y., Goulet, F., Temple, L., Touzard, J.-M. (eds). Éditions Quæ, Versailles.

Sukhwani, V., Nurzaman, A., Kusumawardhani, N., AlHinai, A.-M., Hanyu, L., Shaw, R. (2019). Enhancing food security by institutionalizing collaborative food alliances in urban areas. *Sustainability*, 11(15), 4103.

Tankam, C. (2015). Analyse économique du développement des nouveaux marchés biologiques. PhD Thesis, Université Clermont Auvergne.

Temple, L., Machicou Ndzesop, N., Fongang Fouepe, G.-H., Ndoumbe-Nkeng, M., Mathé, S. (2017). National system of research and innovation in Africa: The case of Cameroon. *Innovations*, 53(2), 41–67.

Thornton, P.-K., Jones, P.-G., Ericksen, P.-J., Challinor, A. (2011). Agriculture and food systems in sub-Saharan Africa in a 4°C+ world. *Philosophical Transactions of the Royal Society A: Mathematical, Physical and Engineering Sciences*, 369(1934), 117–136.

Walther, O.-J., Tenikue, M., Trémolières, M. (2019). Economic performance, gender and social networks in West African food systems. *World Development*, 124(December), 104650.

12

Total Processing of Soy *Glycine max* through Valorization of the Tofu Whey into Cosmetic Products

12.1. Introduction

Natural cosmetics are gaining increasing consumer interest. At the same time, producers are starting to market green-labeled products due to the increased availability of green raw materials and packaging materials (Amber and Fogarassy 2019). The share of natural cosmetic products will further increase in the future market due to a growing motivation for sustainability-related products (Amber 2019). In particular, customers are increasingly convinced by cosmetics based on active ingredients extracted from food products. Indeed, on the one hand, taking care of our external appearance through a good diet is an increasingly attractive lifestyle. On the other hand, scientific evidence of pharmacological activities on the skin of products derived from foodstuffs has become more numerous. In ethical and ecological behavior, cosmetic consumers have a better tendency to use products derived from the processing of food by-products (Faria-Sylvia et al. 2020). Soybean (*Glycine max* (L.) Merr. FABACEAE) is among the plants with the best sources of protein in quality and quantity, and it is considered a low-cost substitute for milk for developing countries (Tang and Liu 2017). Dietary supplementation with these proteins promotes collagen synthesis and proper skin regeneration (Zhang et al. 2020). Administration of soy isoflavones prevents skin aging by conferring better elasticity and reducing wrinkles (Domaszewka-Szostek 2021). When applied topically, soy extract exhibits protective activities against lipid peroxidation and UV

Chapter written by Dinah ANDRIANJAFY, Helga RIM FARASOA and Felamboahangy RASOARAHONA.

radiation (Nagula and Wairkar 2019). Thus, soy fulfills the ambivalent function of acting internally and externally for skin care. Thus, this study has the dual objective of promoting soy from a nutritional point of view by promoting better storability of tofu, a food product derived from soy, and enhancing the value of whey, a co-product of tofu, for the manufacture of cosmetic products. Indeed, tofu, a soy protein concentrate, needs to be refrigerated for its preservation. Its distribution is therefore difficult in a country such as Madagascar, where access to electricity was only 33.7% in 2020 (World Databank 2020), allowing the availability of the cold chain for only a small part of the population.

12.2. Interests in the valorization of soya whey in cosmetics

Degradation of soy whey requires high biochemical and chemical oxygen demands, and several teams are competing in the search for its reuse (Barber et al. 2018). One method for its reuse is to extract its active molecules. Thanks to the development of biotechnology, research on new processes to obtain different products, such as emulsifiers or biofuels, have also been conducted. Nevertheless, all these proposed techniques generate other by-products, and other total transformation solutions would be desirable (Chua and Liu 2019). Thus, the interest in upgrading soy whey into cosmetic products is first ecological.

The other interest is economic. The fruit and vegetable industry forms the most active sector that provides a large amount of by-products. The circular economy is proposed as a solution to reduce waste and generate additional raw materials. The circular economy is a systemic and multiactor approach based on the principle of successive cascades of processes, eliminating the notion of waste and operating on fully renewable resources and energy (Sauve et al. 2016). By-products from the food industry often offer potential for valorization in pharmaceuticals and cosmetics (Osorio et al. 2021).

Finally, soy whey contains interesting biocompounds such as proteins, minerals and especially flavonoids with remarkable antioxidant properties. Zhang et al. (2020) measured the total content of phenolic compounds at 60.87 mg GAE/100 mL (gallic acid equivalent). The antioxidant activity measured by DPPH (2,2-diphenyl-1-picrylhydrazyl) was 65.11 µmol Trolox/100 mL and 14.28 µmol Trolox/100 mL by the FRAP (Ferric Reduction Antioxidant Power) method. Hence, there is interest in its valorization in anti-wrinkle cosmetic products.

12.3. Materials and methods

12.3.1. *Preparation of the tofu*

Tofu was prepared in the laboratory from locally purchased soybean powder. Traditional soy milk coagulation processes were applied with some modifications (Faisal et al. 2014; Nugusu and Gudisa 2016). Briefly, 1.5 L of water was added to 100 g of soybean powder, and the whole was boiled for 50 minutes. The milk was then filtered, and after cooling to 40°C, the coagulant was added. Food grade vinegar with 6% acetic acid was used for coagulation at an amount of 2% vol/vol. After the curd was decanted, the whey was recovered by filtration, and the curd was pressed to obtain a firm paste, tofu. These steps are summarized in Figure 12.1.

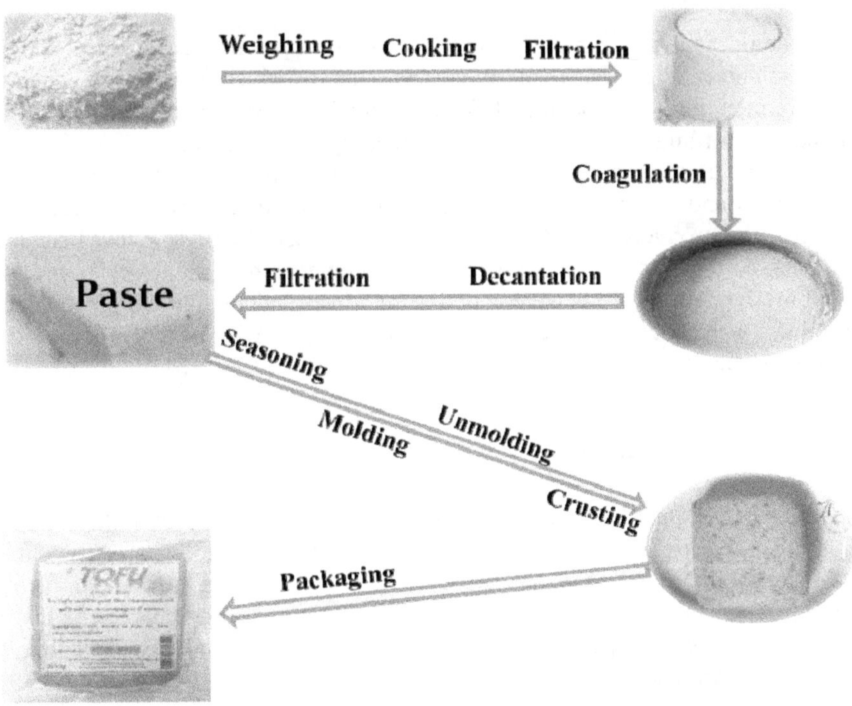

Figure 12.1. *Steps in the production of tofu. For a color version of this figure, see www.iste.co.uk/casadella/agri-innovations.zip*

12.3.2. *Tofu preservation process*

Pink pepper (*Schinus terebinthifolia Raddi* ANACARDIACEAE) was chosen as a preservative. This plant is endowed with antimicrobial and antioxidant activities. The oral LD50 of the fruit was estimated at 5 g/kg, and its use for human consumption does not present any risk (Carvalho et al. 2013). Its antibacterial properties are due, among other things, to its essential oil content, the latter acting by destroying the cell membranes of these bacteria (da Silva Dannenberg et al. 2019). Likewise, its essential oil has antioxidant activity and has been shown to be effective in food preservation (da Silva Dannenberg et al. 2016). The same batch of tofu was divided into untreated control and treated tofus. For the treated tofus, 1%, 1.5% and 2% w/w pink pepper powder was added. To improve the taste, 2.5% w/w table salt was added. After mixing the ingredients, the surface of the tofu was soaked with edible oil, and the whole was put in the oven at 150°C for 15 minutes to form a crust.

12.3.3. *Tofu acceptability test*

A hedonic analysis was performed to evaluate consumer acceptance. Seventy healthy volunteers, aged between 10 and 70 years, of which 54% were women and 46% were men, tested the neutral, 1%, 1.5% and 2% pink pepper tofu. The test was conducted according to the method described by Sharif et al. (2014) with some modifications. The volunteers were asked about the following criteria: appearance, flavor, aroma and texture. Volunteers were also asked to express their preferences by ranking the products from 1 (least liked) to 4 (most liked). To avoid bias in the judgments, no remuneration was given to the volunteers, be it monetary or any other form of motivation. Products were coded using three random digits. Prior to the test, all products were microbiologically tested according to AFNOR V08B standards to ensure their quality.

12.3.4. *Tofu shelf life test*

A sensory evaluation test was conducted to judge the storability of tofu. According to Sharif et al. (2014), for food products, spoilage criteria are sensory first before microbiological quality is affected. Tofus were left at room temperature in the open air and at 4°C, and the evolution of their sensory characteristics was monitored daily. Only tofu appreciated by consumers was tested for shelf life. The untreated tofu was used as a control.

12.3.5. *Preparation of whey before its use in cosmetics*

Given that soy whey degrades very quickly, a preservative of cosmetic products from Clariant has been added to preserve all of its qualities before processing. This preservative is COSMOS certified, and the ingredients are of natural origin. European legislation imposes a limit of 3.1% as a maximum threshold of its use for non-rinseable products. For Japanese legislation, this threshold is 1.33% for all types of products (Clariant 2013). We incorporated it at a rate of 1.25%. To test the effectiveness of this preservative, the batch was divided into two batches: one batch was stored at room temperature, and a second batch was placed in an oven at 45°C (Kirbride et al. 2021). The qualities of both batches were monitored daily with respect to their pH, odors, colors and turbidities. Additionally, their microbiological qualities were monitored according to AFNOR V 08 B standards.

12.3.6. *Development of cosmetic products*

To test the performance of soy whey as an aqueous phase (both active and excipient) in different products, two emulsions were developed. The first emulsion was of the oil-in-water type composed of whey at 86%, vegetable oil at 4%, glycerin at 4%, and vitamin E at 2% and the remainder was composed of thickeners of biological origin of the Emulsan type. A second emulsion was of the water-in-oil type, composed of 10% vegetable oil and 80% whey, but the emulsifier was of the fatty acid polyglyceride type. The vegetable oil used was sunflower oil *Helianthus annuus* L. ASTERACEAE.

12.3.7. *Quality and shelf life tests of cosmetic products*

The organoleptic characteristics studied were color, odor, appearance, texture and penetrability. The pH and density were measured. Their viscosities were measured with a Brookfield viscometer. An accelerated aging test was performed according to the protocol described in Vieira et al. (2009). Forty-eight hours after the preparation of the products, they were conditioned, and this moment was considered the starting time of the test. Two temperatures were chosen: room temperature (22 ± 2°C) and 45°C in the oven. Stability monitoring criteria were established as follows for color, odor and appearance: N for normal, M for modified and IM for intensely modified.

12.3.8. *Efficacy testing of cosmetic products*

Before using the whey, a test to verify its antioxidant activity was performed. The protocol using 2,2-diphenylpicrylhydrazyl (DPPH) was chosen. The whey was diluted to concentrations of 30%, 40%, 50%, 60% and 70% vol/vol in methanol, and the solutions thus prepared were spotted on a silica gel plate using a Pasteur pipette. A solution of 25% DPPH vol/vol in methanol was sprayed on the plate, and a yellow color change demonstrated the existence of antioxidant activity. The same test was performed after the cosmetic products were made to verify whether the activity was maintained or altered during the transformation process. After verification of the good microbiological quality of the cosmetic products developed, 18 female volunteers over 18 years old were tested for safety and efficacy. The products were first subjected to a skin tolerance test using a repeated open application test (ROAT) for five days. The protocol was inspired by the method explained by Bennike et al. (2018), with modifications. The inclusion criteria of the volunteers were as follows: absence of eczema or any other skin disease, not pregnant, not breastfeeding, not exposed to intense UV light during the three weeks before the test and not having applied any other cosmetic products on the elbow crease or face during the three weeks before the test. They were divided into two groups, one group using the oil-in-water emulsion and the other group using the water-in-oil emulsion. If no redness, heat or other signs of irritation were noted, the volunteer could proceed to an efficacy test by applying the product to their face. For two weeks, volunteers were asked to apply the emulsion to clean skin, without rinsing, morning and night. After this period, the volunteers were questioned about the possible existence of an abnormality on the skin, such as a desquamation or a feeling of itching. The evolution of wrinkles or skin hydration was followed by comparing photos before and after the test period.

12.4. Results and interpretations

12.4.1. *Tofu and whey yields*

From 100 g of soybean powder, 120.17 g of tofu was obtained after crouting. This yield is low compared to other improved processes. Lee et al. (2018) obtained a yield of 385.23 g/100 g of soybean meal. One reason for this low yield would be the weak interaction between the coagulant and the proteins. Indeed, even though in their denatured state the proteins are negatively charged, their total charge approaches neutrality due to the presence of protons from the acid addition. There are then fewer hydrophilic–hydrophilic interactions with water, resulting in a lower solubility when using an acid as a coagulant (Sittangang et al. 2020). For whey, approximately 887.50 mL was obtained. This high amount is a major environmental concern, especially for artisanal and small-scale processes (Nugusu and Gudisa 2019). In a developing country, it is relevant to accompany the development of the economy by implementing processes that are not harmful to the environment.

12.4.2. *Sensory analysis of tofu*

Two descriptors were given by the volunteers regarding the appearance of the tofus: characteristic beige color of paste and brown surface color. For the flavor, a spicy sensation was mentioned, as well as a salty taste. For aroma, two characteristics were reported: soy aroma and vinegar aroma. The texture was described by the expressions homogeneous, thick and thick crust.

Figure 12.2 shows the results of the principal component analysis. The graph shows that the three treated tofus are easily distinguishable from the untreated tofu. Only the untreated tofu has a beige paste color, while the other three have a brown color. The level of perception of a vinegar aroma was very low. Soy aroma and spicy flavor were pronounced for the tofu with 1.5% and 2% pink pepper. All three treated tofu samples had a thick consistency and a thick crust, but a homogeneous texture was found for the tofu with 1.5% pink pepper. The graph in Figure 12.3 shows that the tofu with 1% pepper is the most appreciated. At the same time, it is the only sample that shows a strong correlation with a salty taste. However, all four tofu have the same percentage of salt. This could be explained by a strong taste-aroma interaction. Aroma is a prominent factor that can negatively influence taste perception (Gotow et al. 2018). For tofu with 2% pink pepper, the aroma perceived as too spicy deteriorates the taste sensation. Therefore, the preservability test was conducted with 1% pink pepper and 1.5% pink pepper tofu, with a view to marketing-directed research (Filieri 2015).

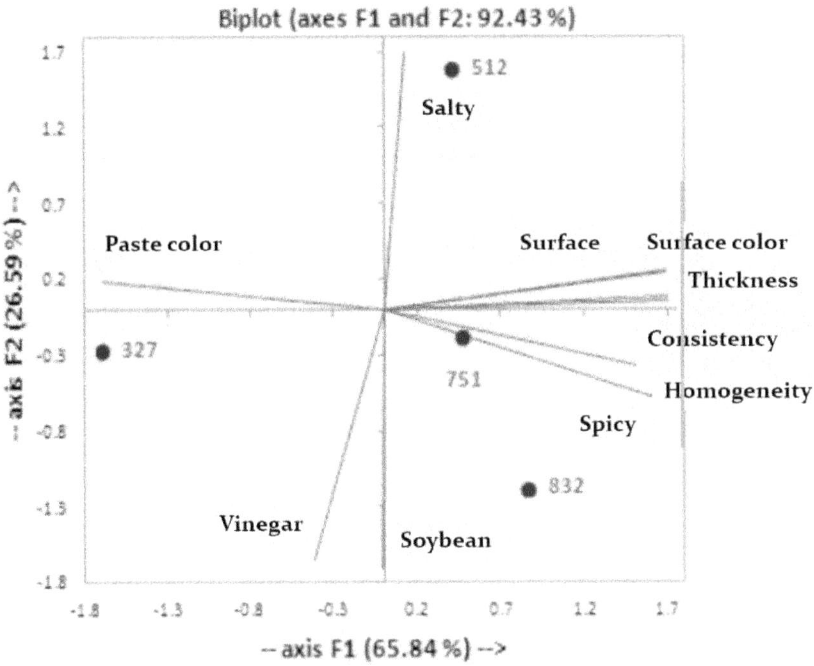

Figure 12.2. *Results of the principal component analysis of tofu. For a color version of this figure, see www.iste.co.uk/casadella/agri-innovations.zip*

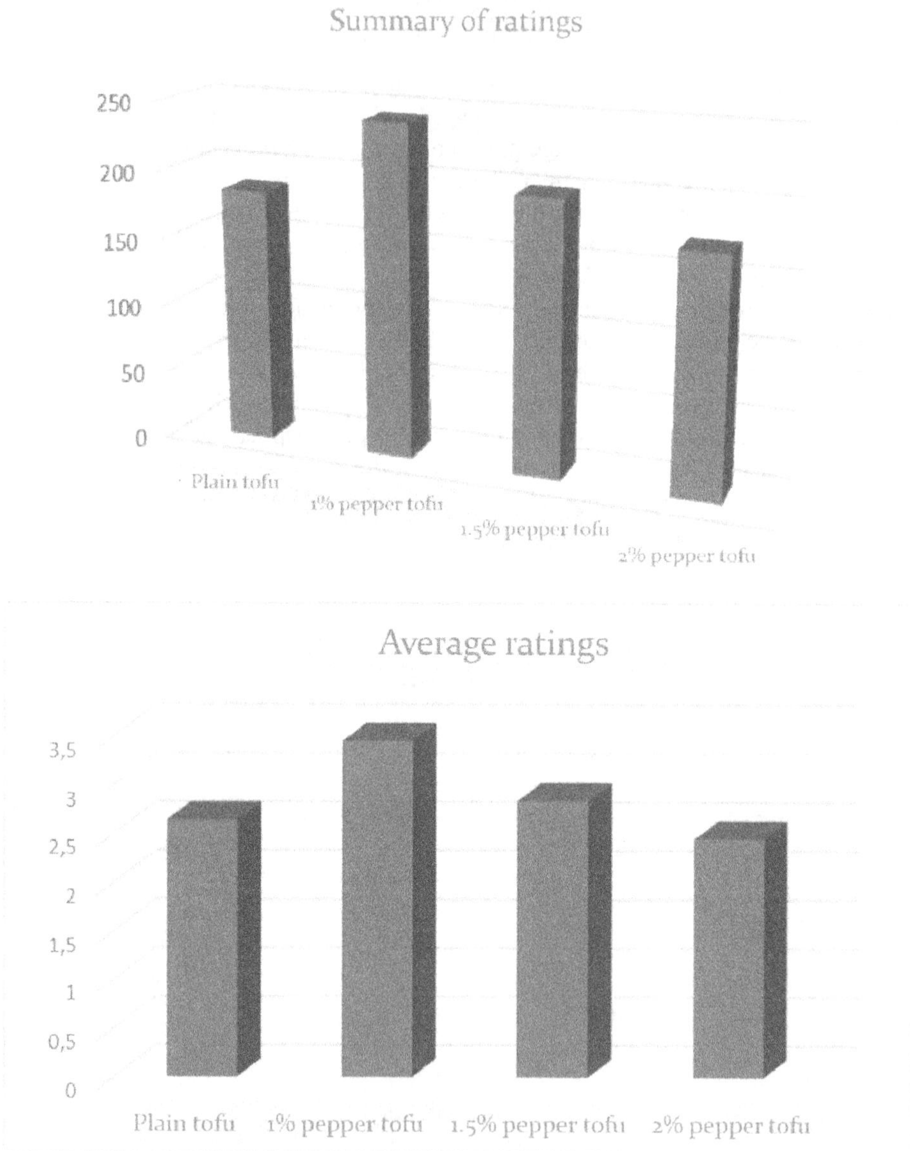

Figure 12.3. *Tofu preference test results. For a color version of this figure, see www.iste.co.uk/casadella/agri-innovations.zip*

12.4.3. *Preservability of processed tofu*

At room temperature, the control tofu could be kept for three days, while under refrigeration, it could be kept for seven days. The 1% tofu was kept for only five days at room temperature and 12 days under refrigeration. At 1.5% pink pepper, the degradation at room temperature started on the eighth day. This would be the equivalent of keeping untreated tofu under refrigeration. At 4°C, tofu with 1.5% pepper could be kept for 15 days. This result is comparable to the shelf life extension obtained by other researchers using 1% tulsi *Ocimum sanctum* L. LAMIACEES (Anbarasu and Vijayalakshmi 2007). It could be inferred that tulsi would perform better than pink pepper in preserving tofu if incorporated at 1%. The signs of spoilage observed were the hardening of the paste and the appearance of a mold on the surface.

12.4.4. *Whey stability*

At a temperature of 45°C, the whey showed a color change to dark yellow with an increase in turbidity after one week. This change stabilized until week 13. Whey stored at room temperature showed no change during the 13 weeks. The pH was stabilized at 5.5. On the other hand, good stability was perceived in terms of odor. The cosmetic preservative was therefore effective. The change in turbidity at 45°C is explained by the reaction of proteins and flavonoids due to heat and does not indicate poor preservation (Vieria et al. 2009). Table 12.1 summarizes the results of microbiological quality control of whey stored at room temperature and stored in the oven after 13 weeks.

12.4.5. *Characterization and stability of the developed cosmetic products*

The oil-in-water emulsion has a milk-like texture, white color and viscosity of 3,474.6 mPa·s. Its relative density is 0.98. Its qualities are comparable to the facial milks sold on the market. The water-in-oil emulsion has a rather unctuous, creamy texture. It is also white in color. Its viscosity is 10,284 mPa·s, and its relative density is 0.97. For both products, no smell of soy is perceived. A light perfume has been added. Both products have a pH of 5.5, which respects the balance of the skin.

During the accelerated aging test, no change in color or odor was perceived for either product. All parameters remained N. Only a slight decrease in viscosity was observed. For the 13 weeks of extreme conditions, it can be concluded that the products are stable and comparable to those on the market.

	Whey stored at 45°C	Whey stored at room temperature	AFNOR V 08 B criteria
Coagulase-positive staphylococci	Absent	Absent	Absent
Yeasts and molds	<10 CFU/g	<10 CFU/g	$<10^2$ CFU/g
Candida albicans	Absent	Absent	Absent
Aspergillus niger	<10 CFU/g	<10 CFU/g	$<10^2$ CFU/g
Microorganism at 30°C	<10 CFU/g	<10 CFU/g	$<10^3$ CFU/g

Table 12.1. *Results of microbiological quality control of whey*

12.4.6. *Effectiveness of the manufactured cosmetic products*

The simple colaration test of DPPH revealed that the small one did not lose its antioxidant virtue after its transformation into cosmetic products. No irritation was observed for the 18 volunteers in the ROAT test. A reduction in fine lines was observed in the older volunteers after daily application of the product, and the volunteers claimed to have more supple skin. Figure 12.4 shows an example of a volunteer photo.

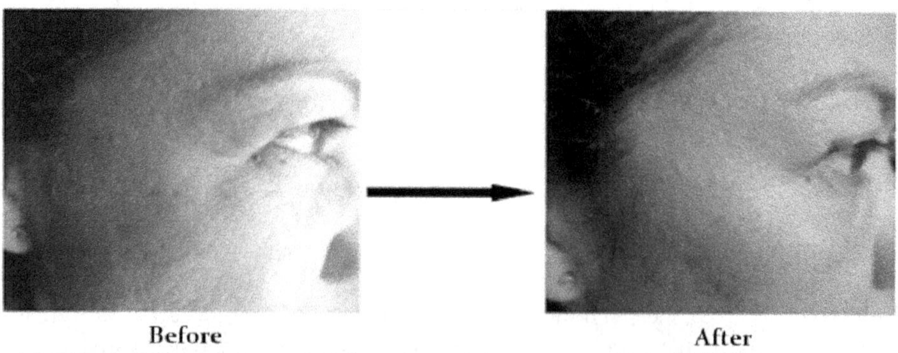

Figure 12.4. *Example of volunteer photos. For a color version of this figure, see www.iste.co.uk/casadella/agri-innovations.zip*

12.4.7. *General discussion*

The possibility of using soy whey in cosmetic products has been demonstrated. The products have good anti-wrinkle properties. Whey can be used both in water-in-oil emulsions and in oil-in-water emulsions. The bioactive molecules in whey offer a wide range of possibilities: anti-aging, sun protection, healing, rehydrating and whitening (Kim et al. 2017). For a greener planet and the need for new raw materials, these findings would pave the way for the development of several other new whey-based cosmetic products. Whey's compatibility with currently available preservatives on the market allows it to be preserved before use in cosmetics. It can be used in its entirety, without further processing, and without generating other by-products. As the main ingredient of the aqueous phase of emulsions, it can be used in large quantities. This reuse meets one of the aspects of the circular economy and the social responsibility of the producing companies. The eco-responsible behavior of producers and consumers is imperative for the current global context (Loizidou and Argyri 2020) and, in particular, so that developing countries can choose their development axes in an appropriate way.

12.5. Conclusion

Soy is a food with multiple virtues, and it has regenerative and strong antioxidant properties for the skin. Its use for cosmetic purposes can be done by oral ingestion or by skin application.

To promote the consumption of soybeans, they have been transformed into tofu, whose taste and preservability have been improved. The addition of 2.5% salt and 1–1.5% pink peppercorns extended the preservation of its quality to seven days at room temperature and 15 days under refrigeration. The volunteers appreciated its taste and texture. The use of vinegar with 6% acetic acid gave a lower yield compared to the use of calcium or magnesium salts.

This research also demonstrated the possibility of full utilization of whey without creating additional by-products by incorporating it into cosmetic products. The processing of soy whey to make it a stable ingredient over time is not difficult. Even without the addition of a fragrance base, no soy odor is disturbed in the final product. In a global context of expansion of the soy industry and to avoid the rejection of high BOD whey, the valorization of the latter in cosmetic products is one of the solutions of choice. The possibility of using it both as an active ingredient and as an excipient offers the possibility of its exploitation in large quantities. Whey can easily be used as a water-in-oil emulsion or an oil-in-water emulsion, with highly stable products, which offers a way to make a very varied range of products, as the properties of soy on the skin are multiple.

The economic expansion of developing countries would necessarily pass by a blooming of industrial activity, particularly in the food-processing sector. It is recommended for these countries to multiply research to favor coproducts rather than rejecting by-products. Environmental impact studies prior to the establishment of industries should not only take into account waste and wastewater treatment processes and socioenvironmental impacts. The potential for full product recovery should also be part of the assessment criteria. Processes with a low percentage of transformation should be disfavored to better apply the circular economy system and preserve sustainable development.

12.6. References

Amberg, N. (2019). Investigating consumer behavior with Natural Cosmetics. *Proceedings of the 9th International Conference on Management*, 1. Szent Istvan University, Hungary.

Amberg, N. and Fogarassy, C. (2019). Green consumer behavior in the cosmetics market. *Resources*, 8, 137.

Anbarasu, K. and Vijayalakshmi, G. (2007). Improved shelf life of protein-rich tofu using *Ocimum sanctum* (tulsi) extracts to benefit Indian rural population. *Journal of Food Science*, 72, 8.

Barber, S.T., Yin, J., Draper, K., Trabold, T.A. (2018). Closing nutrient cycles with biochar – From filtration to fertilizer. *Journal of Cleaner Production*, 197(1), 1597–1606.

Bennike, N.H., Palangi, L., Christensson, J.B., Nilsson, U., Zachariae, C., Johansen, J.D., Hagvall, L. (2019). Allergic contact dermatitis caused by hydroperoxides of limonene and dose-response relationship – A repeated open application test (ROAT) study. *Contact Dermatitis*, 80(4), 208–216.

Carvalho, M.G., Melo, A.G.N., Aragão, C.F.S., Raffin, F.N., Moura, T.F.A.L. (2013). *Schinus terebinthifolius* Raddi: Chemical composition, biological properties and toxicity. *Revista Brasileira de Plantas Medicinais*, 15, 158–169.

Chua, J.Y. and Liu, S.Q. (2019). Soy whey: More than just wastewater from tofu and soy protein isolate industry. *Trends in Food Science & Technology*, 91, 24–32.

Clariant (2013). Nipaguard Product Fact Sheet [Online]. Available at: https://www.clariant.com/en/Solutions/Products/2014/05/02/12/33/Nipaguard-SCE.

Domaszewka-Szostek, A., Puzianowska-Kuźnicka, M., Kuryłowicz, A. (2021). Flavonoids in skin senescence prevention and treatment. *International Journal of Molecular Sciences*, 22, 6814.

Faisal, M., Mulana, F., Daimon, H., Machdar, I. (2014). Potential renewable energy from tofu processing waste in Banda Aceh City, Indonesia. *Asian Journal of Chemistry*, 26(19), 6601–6604.

Faria-Sylvia, C., Ascenso, A., Costa, A., Marto, J., Carvalheiro, M., Ribeiro, H.M., Simões, S. (2020). Feeding the skin: A new trend in food and cosmetics convergence. *Trends in Food Sciences and Technology*, 95, 21–32.

Filieri, R. (2015). From market-driving to market-driven: An analysis of Benetton's strategy change and its implications for long-term performance. *Marketing Intelligence and Planning*, 3, 33.

Gotow, N., Skrandies, W., Kobayashi, T., Kobayakawa, T. (2018). Familiarity and retronasal aroma alter food perception. *Chemosensory Perception*, 11, 77–94.

Kim, J.-E., Kang, Y.-G., Park, J.S., Lim, T.-G., Lee, K.W. (2017). Review of soybean phytochemicals and their bioactive properties relevant for skin health. *Journal of Food and Nutrition Research*, 5(11), 852–858.

Kirkbride, L., Humphries, L., Kozielska, P., Curtis, H. (2021). Designing a suitable stability protocol in the face of a changing retail landscape. *Cosmetics*, 8(3), 64.

Lee, K.Y., Rahman, M.S., Kim, A.N., Gul, K., Kang, S.W., Chun, J., Kerr, W.L., Choi, S.G. (2019). Quality characteristics and storage stability of low-fat tofu prepared with defatted soy flours treated by supercritical-CO_2 and hexane. *Lwt Food Science & Technology*, 100, 237–243.

Loizidou, A. and Argyri, P. (2020). Circular economy and innovative business. *Open Schools Journals for Open Science*, 3, 3.

Nagula, R.L. and Wairkar, S. (2019). Recent advances in topical delivery of flavonoids, a review. *Journal of Controlled Release*, 296, 190–201.

Nugroho, G.S.F., Sulistyaningrum, R., Melania, R.P., Handayani, W. (2019). Environmental analysis of tofu production in the context of cleaner production: Case study of tofu household industries in Salatiga, Indonesia. *Journal of Environmental Science and Sustainable Development*, 2(2), 127–138.

Nugusu, Y. and Gudisa, A. (2019). Evaluation of coagulants on soy cheese making efficiency. *International Journal of Trend in Research and Development*, 3, 1.

Osorio, L.R.D., Flórez-López, E., Grande-Tovar, C.D. (2021). The potential of selected agri-food loss and waste to contribute to a circular economy: Applications in the food, cosmetic and pharmaceutical industries. *Molecules*, 26, 515.

Sauvé, S., Normandin, D., McDonald, M. (2016). *L'économie circulaire : une transition incontournable*. Les Presses de l'Université de Montréal.

Sharif, M.K., Butt, M.S., Sharif, H.R., Nasir, M. (2014). Sensory evaluation and consumer acceptability. In *Handbook of Food Science and Technology*, Butt, M.S. and Zahor, T. (eds). UAF Press, Faisalabad.

da Silva Dannenberg, G., Funck, G.D., Mattei, F.J., da Silva, W.P., Fiorentini, Â.M. (2016). Antimicrobial and antioxidant activity of essential oil from pink pepper tree (Schinus terebinthifolius Raddi) in vitro and in cheese experimentally contaminated with Listeria monocytogenes. *Innovative Food Science & Emerging Technologies*, 36, 120–127.

da Silva Dannenberg, G., Funck, G.D., da Silva, W.P., Fiorentini, Â.M. (2019). Essential oil from pink pepper (Schinus terebinthifolius Raddi): Chemical composition, antibacterial activity and mechanism of action. *Food Control*, 95, 115–120.

Sitanggang, A.B., Alexander, R., Budijanto, S. (2020). The utilization of bilimbi (*Averrhoa bilimbi*) and lime (*Citrus aurantifolia*) juices as natural acid coagulants for tofu production. *Food Sciences and Technology*, 57(12), 4660–4670.

Tang, X. and Liu, J. (2017). A comparative study of partial replacement of wheat flour with whey and soy protein on rheological properties of dough and cookie quality. *Journal of Food Quality*, 1–10. doi: 10.1155/2017/2618020.

Vieria, R.P., Fernandes, A.R., Kaneko, T.M., Consiglieri, V.O., Pinto, C.A.S.O., Pereira, C.S.C., Baby, A.R., Velasco, M.V.R. (2009). Physical and physicochemical stability evaluation of cosmetic formulations containing soybean extract fermented by *Bifdobacterium animalis*. *Brazilian Journal of Pharmaceutical Sciences*, 45, 3.

World Databank (2020). Indicator for Madagascar, access to electricity [Online]. Available at: https://data.worldbank.org/indicator/EG.ELC.ACCS.ZS?locations=MG.

Zhang, J., Fu, X., Li, W., Li, H., Ying, Z., Liu, X., Yin, L. (2020). Enhancement of nutritional soy protein and peptide supplementation on skin repair in rats. *Journal of Functional Foods*, 75, 104231.

Zhao, C.-C., Kim, P.-H., Eun, J.-B. (2020). Influence of high-intensity ultrasound application on the physicochemical properties, isoflavone composition, and antioxidant activity of tofu whey. *LWT-Food Science and Technology*, 117, 108618.

List of Authors

Ainazo ANDRIAMANANTENA
École Doctorale Ecosystèmes Naturels (EDEN)
Université de Mahajanga (UMG)
Madagascar

Fenozo ANDRIAMANANTENA
École Doctorale Ecosystèmes Naturels (EDEN)
Université de Mahajanga (UMG)
Madagascar

Simon Georges ANDRIANASETRA
Information et Documentation Scientifique et Technique
CIDST
Madagascar

Tiantsoa ANDRIANASETRA
École Doctorale Ecosystèmes Naturels (EDEN)
Université de Mahajanga (UMG)
Madagascar

Dinah ANDRIANJAFY
Laboratoire HOMEOPHARMA
Sambaina
Antananarivo
Madagascar

Blaise Waly BASSE
Department of Economics and Management
Université Assane Seck de Ziguinchor
Senegal

Michaël BEFINOANA
École Doctorale Ecosystèmes Naturels (EDEN)
Université de Mahajanga (UMG)
Madagascar

Néfissa BEN MAHMOUD
Laboratoire Prisme
Institut Supérieur de Gestion de Tunis
Université de Tunis
Tunisia

Yasmine BOUGHZALA
Laboratoire Prisme
Institut Supérieur de Gestion de Tunis
Université de Tunis
Tunisia

Paul BOUVIER-PATRON
CERGAM
Université Aix-Toulon
France

Christophe BRING
Department of Geography
Nature Protection and Sustainable Development
Université de Ngaoundéré
Cameroon

Vanessa CASADELLA
LEFMI
Université Picardie Jules Verne
Amiens
and
Reasearch Network on Innovation
Paris
France

Gérard DE LA PAIX BAYIHA
Université de Yaoundé II
Cameroon

Omar DIOP
Department of Economics and Management
Université Assane Seck de Ziguinchor
Senegal

Eric Joël FOFIRI NZOSSIÉ
Department of Geography
UMR 201 Development & Societies
IRD-IEDES
Université de Ngaoundéré
Cameroon

Jean-Luc HORNICK
Department of Veterinary Animal Resource Management
Faculty of Veterinary Medicine
FARAH Center
Université de Liège
Belgium

Albertine Bayompe KABOU
Cheikh Anta Diop University
Dakar
Senegal

Son Thi Kim LE
Centre de recherche sur l'Innovation et les Stratégies Industrielles (ISI)
Laboratoire de Recherche sur l'Industrie et l'Innovation
Université du Littoral
Côte d'Opale
Dunkirk
and
Reasearch Network on Innovation
Paris
France

Cyrille MAHAROMBAKA
École Doctorale Ecosystèmes Naturels (EDEN)
Université de Mahajanga (UMG)
Madagascar

Syndhia MATHÉ
CIRAD
UMR Innovation
France
and
Science and Technology Policy Research Institute
CSIR-STEPRI
Accra
Ghana
and
INNOVATION Unit
Institut Agro Montpellier
CIRAD
INRAE
University of Montpellier
France

Souleymane MBAYE
Department of Economics and Management
Université Assane Seck de Ziguinchor
Senegal

Babacar NDIAYE
Department of Economics and Management
Amadou Mahtar Mbow University
Dakar
Senegal

Phuong Anh Thi NGUYEN
Faculty of Political Economy
University of Economics and Business
Vietnam National University
Ho Chi Minh
Vietnam

To The NGUYEN
Faculty of Political Economy
University of Economics and Business
Vietnam National University
Ho Chi Minh
Vietnam

Darryl Neil NITCHEU WAKPONOU
Department of Geography
Université de Ngaoundéré
Cameroon

Antoine Zafera RABESA
École Doctorale Ecosystèmes Naturels (EDEN)
Université de Mahajanga (UMG)
Madagascar

Liva Jackson RAHARINAIVO
École Doctorale Ecosystèmes Naturels (EDEN)
Université de Mahajanga (UMG)
Madagascar

Vololoniaina RAMAHATAFANDRY
École Doctorale Ecosystèmes Naturels (EDEN)
Université de Mahajanga (UMG)
Madagascar

Auguste Botovao RAMIANDRISOA
École Doctorale Ecosystèmes Naturels (EDEN)
Université de Mahajanga (UMG)
Madagascar

Hery Lisy Tiana RANARIJAONA
École Doctorale Ecosystèmes Naturels (EDEN)
Université de Mahajanga (UMG)
Madagascar

Felamboahangy RASOARAHONA
Higher School of Agronomic Sciences
University of Antananarivo
Madagascar

Helga RIM FARASOA
Chemistry of Natural Substances Department
Faculty of Science
University of Antananarivo
Madagascar

Natewindé SAWADOGO
Université Ouagadougou II
Burkina Faso

Eveline M.F.W. SAWADOGO/ COMPAORÉ
Institut de l'Environnement et des Recherches Agricoles
Ouagadougou
Burkina Faso

Ludovic TEMPLE
CIRAD
UMR Innovation
Montpellier
France

Sylvana TOMBOANONA
École Doctorale Ecosystèmes Naturels (EDEN)
Université de Mahajanga (UMG)
Madagascar

Chéri Christian TOTONDRABESA
École Doctorale Ecosystèmes Naturels (EDEN)
Université de Mahajanga (UMG)
Madagascar

Dimitri UZUNIDIS
Reasearch Network on Innovation
Paris
France

Index

G, I

K, L, M

N, P, Q

R, S

Other titles from

iSTE

in

Innovation, Entrepreneurship and Management

2023

SAULAIS Pierre
Knowledge and Ideation: Inventive Knowledge Analysis for Ideation Stimulation (Innovation and Technology Set – Volume 17)

2022

AOUINAÏT Camille
Open Innovation Strategies (Smart Innovation Set – Volume 39)

BOUCHÉ Geneviève
Productive Economy, Contributory Economy: Governance Tools for the Third Millennium (Innovation and Technology Set – Volume 15)

BRUYÈRE Christelle
Caring Management in Health Organizations: A Lever for Crisis Management (Health and Innovation Set – Volume 3)

HELLER David
Valuation of the Liability Structure by Real Options (Modern Finance, Management Innovation and Economic Growth Set – Volume 5)

MATHIEU Valérie
A Customer-oriented Manager for B2B Services: Principles and Implementation

MORALES Lucía, DZEVER Sam, TAYLOR Robert
Asia-Europe Industrial Connectivity in Times of Crisis (Innovation and Technology Set – Volume 16)

NOËL Florent, SCHMIDT Géraldine
Employability and Industrial Mutations: Between Individual Trajectories and Organizational Strategic Planning (Technological Changes and Human Resources Set – Volume 4)

DE SAINT JULIEN Odile
The Innovation Ecosystem as a Source of Value Creation: A Value Creation Lever for Open Innovation (Diverse and Global Perspectives on Value Creation Set – Volume 4)

SALOFF-COSTE Michel
Innovation Ecosystems: The Future of Civilizations and the Civilization of the Future (Innovation and Technology Set – Volume 14)

VAYRE Emilie
Digitalization of Work: New Spaces and New Working Times (Technological Changes and Human Resources Set – Volume 5)

ZAFEIRIS Konstantinos N, SKIADIS Christos H, DIMOTIKALIS Yannis, KARAGRIGORIOU Alex, KARAGRIGORIOU-VONTA Christina
Data Analysis and Related Applications 1: Computational, Algorithmic and Applied Economic Data Analysis (Big Data, Artificial Intelligence and Data Analysis Set – Volume 9)
Data Analysis and Related Applications 2: Multivariate, Health and Demographic Data Analysis (Big Data, Artificial Intelligence and Data Analysis Set – Volume 10)

2021

ARCADE Jacques
Strategic Engineering (Innovation and Technology Set – Volume 11)

BÉRANGER Jérôme, RIZOULIÈRES Roland
The Digital Revolution in Health (Health and Innovation Set – Volume 2)

BOBILLIER CHAUMON Marc-Eric
Digital Transformations in the Challenge of Activity and Work: Understanding and Supporting Technological Changes (Technological Changes and Human Resources Set – Volume 3)

BUCLET Nicolas
Territorial Ecology and Socio-ecological Transition (Smart Innovation Set – Volume 34)

DIMOTIKALIS Yannis, KARAGRIGORIOU Alex, PARPOULA Christina, SKIADIS Christos H
Applied Modeling Techniques and Data Analysis 1: Computational Data Analysis Methods and Tools (Big Data, Artificial Intelligence and Data Analysis Set - Volume 7)
Applied Modeling Techniques and Data Analysis 2: Financial, Demographic, Stochastic and Statistical Models and Methods (Big Data, Artificial Intelligence and Data Analysis Set – Volume 8)

DISPAS Christophe, KAYANAKIS Georges, SERVEL Nicolas, STRIUKOVA Ludmila
Innovation and Financial Markets (Innovation between Risk and Reward Set – Volume 7)

ENJOLRAS Manon
Innovation and Export: The Joint Challenge of the Small Company (Smart Innovation Set – Volume 37)

FLEURY Sylvain, RICHIR Simon
Immersive Technologies to Accelerate Innovation: How Virtual and Augmented Reality Enables the Co-Creation of Concepts (Smart Innovation Set – Volume 38)

GIORGINI Pierre
The Contributory Revolution (Innovation and Technology Set – Volume 13)

GOGLIN Christian
Emotions and Values in Equity Crowdfunding Investment Choices 2: Modeling and Empirical Study

GRENIER Corinne, OIRY Ewan
Altering Frontiers: Organizational Innovations in Healthcare (Health and Innovation Set – Volume 1)

GUERRIER Claudine
Security and Its Challenges in the 21st Century (Innovation and Technology Set – Volume 12)

HELLER David
Performance of Valuation Methods in Financial Transactions (Modern Finance, Management Innovation and Economic Growth Set – Volume 4)

LEHMANN Paul-Jacques
Liberalism and Capitalism Today

SOULÉ Bastien, HALLÉ Julie, VIGNAL Bénédicte, BOUTROY Éric, NIER Olivier
Innovation in Sport: Innovation Trajectories and Process Optimization (Smart Innovation Set – Volume 35)

UZUNIDIS Dimitri, KASMI Fedoua, ADATTO Laurent
Innovation Economics, Engineering and Management Handbook 1: Main Themes
Innovation Economics, Engineering and Management Handbook 2: Special Themes

VALLIER Estelle
Innovation in Clusters: Science–Industry Relationships in the Face of Forced Advancement (Smart Innovation Set – Volume 36)

2020

ACH Yves-Alain, RMADI-SAÏD Sandra
Financial Information and Brand Value: Reflections, Challenges and Limitations

ANDREOSSO-O'CALLAGHAN Bernadette, DZEVER Sam, JAUSSAUD Jacques, TAYLOR Robert
Sustainable Development and Energy Transition in Europe and Asia (Innovation and Technology Set – Volume 9)

BEN SLIMANE Sonia, M'HENNI Hatem
Entrepreneurship and Development: Realities and Future Prospects (Smart Innovation Set – Volume 30)

CHOUTEAU Marianne, FOREST Joëlle, NGUYEN Céline
Innovation for Society: The P.S.I. Approach (Smart Innovation Set – Volume 28)

CORON Clotilde
Quantifying Human Resources: Uses and Analysis (Technological Changes and Human Resources Set – Volume 2)

CORON Clotilde, GILBERT Patrick
Technological Change (Technological Changes and Human Resources Set – Volume 1)

CERDIN Jean-Luc, PERETTI Jean-Marie
The Success of Apprenticeships: Views of Stakeholders on Training and Learning (Human Resources Management Set – Volume 3)

DELCHET-COCHET Karen
Circular Economy: From Waste Reduction to Value Creation (Economic Growth Set – Volume 2)

DIDAY Edwin, GUAN Rong, SAPORTA Gilbert, WANG Huiwen
Advances in Data Science (Big Data, Artificial Intelligence and Data Analysis Set – Volume 4)

DOS SANTOS PAULINO Victor
Innovation Trends in the Space Industry (Smart Innovation Set – Volume 25)

GASMI Nacer
Corporate Innovation Strategies: Corporate Social Responsibility and Shared Value Creation (Smart Innovation Set – Volume 33)

GOGLIN Christian
Emotions and Values in Equity Crowdfunding Investment Choices 1: Transdisciplinary Theoretical Approach

GUILHON Bernard
Venture Capital and the Financing of Innovation (Innovation Between Risk and Reward Set – Volume 6)

LATOUCHE Pascal
Open Innovation: Human Set-up (Innovation and Technology Set – Volume 10)

LIMA Marcos
Entrepreneurship and Innovation Education: Frameworks and Tools (Smart Innovation Set – Volume 32)

MACHADO Carolina, DAVIM J. Paulo
Sustainable Management for Managers and Engineers

MAKRIDES Andreas, KARAGRIGORIOU Alex, SKIADAS Christos H.
Data Analysis and Applications 3: Computational, Classification, Financial, Statistical and Stochastic Methods (Big Data, Artificial Intelligence and Data Analysis Set – Volume 5)
Data Analysis and Applications 4: Financial Data Analysis and Methods (Big Data, Artificial Intelligence and Data Analysis Set – Volume 6)

MASSOTTE Pierre, CORSI Patrick
Complex Decision-Making in Economy and Finance

MEUNIER François-Xavier
Dual Innovation Systems: Concepts, Tools and Methods (Smart Innovation Set – Volume 31)

MICHAUD Thomas
Science Fiction and Innovation Design (Innovation in Engineering and Technology Set – Volume 6)

MONINO Jean-Louis
Data Control: Major Challenge for the Digital Society (Smart Innovation Set – Volume 29)

MORLAT Clément
Sustainable Productive System: Eco-development versus Sustainable Development (Smart Innovation Set – Volume 26)

SAULAIS Pierre, ERMINE Jean-Louis
Knowledge Management in Innovative Companies 2: Understanding and Deploying a KM Plan within a Learning Organization
(Smart Innovation Set – Volume 27)

2019

AMENDOLA Mario, GAFFARD Jean-Luc
Disorder and Public Concern Around Globalization

BARBAROUX Pierre
Disruptive Technology and Defence Innovation Ecosystems
(Innovation in Engineering and Technology Set – Volume 5)

DOU Henri, JUILLET Alain, CLERC Philippe
Strategic Intelligence for the Future 1: A New Strategic and Operational Approach
Strategic Intelligence for the Future 2: A New Information Function Approach

FRIKHA Azza
Measurement in Marketing: Operationalization of Latent Constructs

FRIMOUSSE Soufyane
Innovation and Agility in the Digital Age
(Human Resources Management Set – Volume 2)

GAY Claudine, SZOSTAK Bérangère L.
Innovation and Creativity in SMEs: Challenges, Evolutions and Prospects
(Smart Innovation Set – Volume 21)

GORIA Stéphane, HUMBERT Pierre, ROUSSEL Benoît
Information, Knowledge and Agile Creativity
(Smart Innovation Set – Volume 22)

HELLER David
Investment Decision-making Using Optional Models
(Economic Growth Set – Volume 2)

HELLER David, DE CHADIRAC Sylvain, HALAOUI Lana, JOUVET Camille
The Emergence of Start-ups
(Economic Growth Set – Volume 1)

HÉRAUD Jean-Alain, KERR Fiona, BURGER-HELMCHEN Thierry
Creative Management of Complex Systems
(Smart Innovation Set – Volume 19)

LATOUCHE Pascal
Open Innovation: Corporate Incubator
(Innovation and Technology Set – Volume 7)

LEHMANN Paul-Jacques
The Future of the Euro Currency

LEIGNEL Jean-Louis, MÉNAGER Emmanuel, YABLONSKY Serge
Sustainable Enterprise Performance: A Comprehensive Evaluation Method

LIÈVRE Pascal, AUBRY Monique, GAREL Gilles
Management of Extreme Situations: From Polar Expeditions to Exploration-Oriented Organizations

MILLOT Michel
Embarrassment of Product Choices 2: Towards a Society of Well-being

N'GOALA Gilles, PEZ-PÉRARD Virginie, PRIM-ALLAZ Isabelle
Augmented Customer Strategy: CRM in the Digital Age

NIKOLOVA Blagovesta
The RRI Challenge: Responsibilization in a State of Tension with Market Regulation
(Innovation and Responsibility Set – Volume 3)

PELLEGRIN-BOUCHER Estelle, ROY Pierre
Innovation in the Cultural and Creative Industries
(Innovation and Technology Set – Volume 8)

PRIOLON Joël
Financial Markets for Commodities

QUINIOU Matthieu
Blockchain: The Advent of Disintermediation

RAVIX Joël-Thomas, DESCHAMPS Marc
Innovation and Industrial Policies
(Innovation between Risk and Reward Set – Volume 5)

ROGER Alain, VINOT Didier
Skills Management: New Applications, New Questions
(Human Resources Management Set – Volume 1)

SAULAIS Pierre, ERMINE Jean-Louis
Knowledge Management in Innovative Companies 1: Understanding and Deploying a KM Plan within a Learning Organization
(Smart Innovation Set – Volume 23)

SERVAJEAN-HILST Romaric
Co-innovation Dynamics: The Management of Client-Supplier Interactions for Open Innovation
(Smart Innovation Set – Volume 20)

SKIADAS Christos H., BOZEMAN James R.
Data Analysis and Applications 1: Clustering and Regression, Modeling-estimating, Forecasting and Data Mining
(Big Data, Artificial Intelligence and Data Analysis Set – Volume 2)
Data Analysis and Applications 2: Utilization of Results in Europe and Other Topics
(Big Data, Artificial Intelligence and Data Analysis Set – Volume 3)

UZUNIDIS Dimitri
Systemic Innovation: Entrepreneurial Strategies and Market Dynamics

VIGEZZI Michel
World Industrialization: Shared Inventions, Competitive Innovations and Social Dynamics
(Smart Innovation Set – Volume 24)

2018

BURKHARDT Kirsten
Private Equity Firms: Their Role in the Formation of Strategic Alliances

CALLENS Stéphane
Creative Globalization
(Smart Innovation Set – Volume 16)

CASADELLA Vanessa
Innovation Systems in Emerging Economies: MINT – Mexico, Indonesia, Nigeria, Turkey
(Smart Innovation Set – Volume 18)

CHOUTEAU Marianne, FOREST Joëlle, NGUYEN Céline
Science, Technology and Innovation Culture
(Innovation in Engineering and Technology Set – Volume 3)

CORLOSQUET-HABART Marine, JANSSEN Jacques
Big Data for Insurance Companies
(Big Data, Artificial Intelligence and Data Analysis Set – Volume 1)

CROS Françoise
Innovation and Society
(Smart Innovation Set – Volume 15)

DEBREF Romain
Environmental Innovation and Ecodesign: Certainties and Controversies
(Smart Innovation Set – Volume 17)

DOMINGUEZ Noémie
SME Internationalization Strategies: Innovation to Conquer New Markets

ERMINE Jean-Louis
Knowledge Management: The Creative Loop
(Innovation and Technology Set – Volume 5)

GILBERT Patrick, BOBADILLA Natalia, GASTALDI Lise,
LE BOULAIRE Martine, LELEBINA Olga
Innovation, Research and Development Management

IBRAHIMI Mohammed
Mergers & Acquisitions: Theory, Strategy, Finance

LEMAÎTRE Denis
Training Engineers for Innovation

LÉVY Aldo, BEN BOUHENI Faten, AMMI Chantal
Financial Management: USGAAP and IFRS Standards
(Innovation and Technology Set – Volume 6)

MILLOT Michel
Embarrassment of Product Choices 1: How to Consume Differently

PANSERA Mario, OWEN Richard
Innovation and Development: The Politics at the Bottom of the Pyramid
(Innovation and Responsibility Set – Volume 2)

RICHEZ Yves
Corporate Talent Detection and Development

SACHETTI Philippe, ZUPPINGER Thibaud
New Technologies and Branding
(Innovation and Technology Set – Volume 4)

SAMIER Henri
Intuition, Creativity, Innovation

TEMPLE Ludovic, COMPAORÉ SAWADOGO Eveline M.F.W.
Innovation Processes in Agro-Ecological Transitions in Developing Countries
(Innovation in Engineering and Technology Set – Volume 2)

UZUNIDIS Dimitri
Collective Innovation Processes: Principles and Practices
(Innovation in Engineering and Technology Set – Volume 4)

VAN HOOREBEKE Delphine
The Management of Living Beings or Emo-management

2017

AÏT-EL-HADJ Smaïl
The Ongoing Technological System
(Smart Innovation Set – Volume 11)

BAUDRY Marc, DUMONT Béatrice
Patents: Prompting or Restricting Innovation?
(Smart Innovation Set – Volume 12)

BÉRARD Céline, TEYSSIER Christine
Risk Management: Lever for SME Development and Stakeholder Value Creation

CHALENÇON Ludivine
Location Strategies and Value Creation of International Mergers and Acquisitions

CHAUVEL Danièle, BORZILLO Stefano
The Innovative Company: An Ill-defined Object
(Innovation between Risk and Reward Set – Volume 1)

CORSI Patrick
Going Past Limits To Growth

D'ANDRIA Aude, GABARRET Inés
Building 21st Century Entrepreneurship
(Innovation and Technology Set – Volume 2)

DAIDJ Nabyla
Cooperation, Coopetition and Innovation
(Innovation and Technology Set – Volume 3)

FERNEZ-WALCH Sandrine
The Multiple Facets of Innovation Project Management
(Innovation between Risk and Reward Set – Volume 4)

FOREST Joëlle
Creative Rationality and Innovation
(Smart Innovation Set – Volume 14)

GUILHON Bernard
Innovation and Production Ecosystems
(Innovation between Risk and Reward Set – Volume 2)

HAMMOUDI Abdelhakim, DAIDJ Nabyla
Game Theory Approach to Managerial Strategies and Value Creation
(Diverse and Global Perspectives on Value Creation Set – Volume 3)

LALLEMENT Rémi
Intellectual Property and Innovation Protection: New Practices and New Policy Issues
(Innovation between Risk and Reward Set – Volume 3)

LAPERCHE Blandine
Enterprise Knowledge Capital
(Smart Innovation Set – Volume 13)

LEBERT Didier, EL YOUNSI Hafida
International Specialization Dynamics
(Smart Innovation Set – Volume 9)

MAESSCHALCK Marc
Reflexive Governance for Research and Innovative Knowledge
(Responsible Research and Innovation Set – Volume 6)

MASSOTTE Pierre
Ethics in Social Networking and Business 1: Theory, Practice and Current Recommendations
Ethics in Social Networking and Business 2: The Future and Changing Paradigms

MASSOTTE Pierre, CORSI Patrick
Smart Decisions in Complex Systems

MEDINA Mercedes, HERRERO Mónica, URGELLÉS Alicia
Current and Emerging Issues in the Audiovisual Industry
(Diverse and Global Perspectives on Value Creation Set – Volume 1)

MICHAUD Thomas
Innovation, Between Science and Science Fiction
(Smart Innovation Set – Volume 10)

PELLÉ Sophie
Business, Innovation and Responsibility
(Responsible Research and Innovation Set – Volume 7)

SAVIGNAC Emmanuelle
The Gamification of Work: The Use of Games in the Workplace

SUGAHARA Satoshi, DAIDJ Nabyla, USHIO Sumitaka
Value Creation in Management Accounting and Strategic Management: An Integrated Approach
(Diverse and Global Perspectives on Value Creation Set –Volume 2)

UZUNIDIS Dimitri, SAULAIS Pierre
Innovation Engines: Entrepreneurs and Enterprises in a Turbulent World
(Innovation in Engineering and Technology Set – Volume 1)

2016

BARBAROUX Pierre, ATTOUR Amel, SCHENK Eric
Knowledge Management and Innovation
(Smart Innovation Set – Volume 6)

BEN BOUHENI Faten, AMMI Chantal, LEVY Aldo
Banking Governance, Performance And Risk-Taking: Conventional Banks Vs Islamic Banks

BOUTILLIER Sophie, CARRÉ Denis, LEVRATTO Nadine
Entrepreneurial Ecosystems (Smart Innovation Set – Volume 2)

BOUTILLIER Sophie, UZUNIDIS Dimitri
The Entrepreneur (Smart Innovation Set – Volume 8)

BOUVARD Patricia, SUZANNE Hervé
Collective Intelligence Development in Business

GALLAUD Delphine, LAPERCHE Blandine
Circular Economy, Industrial Ecology and Short Supply Chains
(Smart Innovation Set – Volume 4)

GUERRIER Claudine
Security and Privacy in the Digital Era
(Innovation and Technology Set – Volume 1)

MEGHOUAR Hicham
Corporate Takeover Targets

MONINO Jean-Louis, SEDKAOUI Soraya
Big Data, Open Data and Data Development
(Smart Innovation Set – Volume 3)

MOREL Laure, LE ROUX Serge
Fab Labs: Innovative User
(Smart Innovation Set – Volume 5)

PICARD Fabienne, TANGUY Corinne
Innovations and Techno-ecological Transition
(Smart Innovation Set – Volume 7)

2015

CASADELLA Vanessa, LIU Zeting, DIMITRI Uzunidis
Innovation Capabilities and Economic Development in Open Economies
(Smart Innovation Set – Volume 1)

CORSI Patrick, MORIN Dominique
Sequencing Apple's DNA

CORSI Patrick, NEAU Erwan
Innovation Capability Maturity Model

FAIVRE-TAVIGNOT Bénédicte
Social Business and Base of the Pyramid

GODÉ Cécile
Team Coordination in Extreme Environments

MAILLARD Pierre
Competitive Quality and Innovation

MASSOTTE Pierre, CORSI Patrick
Operationalizing Sustainability

MASSOTTE Pierre, CORSI Patrick
Sustainability Calling

2014

DUBÉ Jean, LEGROS Diègo
Spatial Econometrics Using Microdata

LESCA Humbert, LESCA Nicolas
Strategic Decisions and Weak Signals

2013

HABART-CORLOSQUET Marine, JANSSEN Jacques, MANCA Raimondo
VaR Methodology for Non-Gaussian Finance

2012

DAL PONT Jean-Pierre
Process Engineering and Industrial Management

MAILLARD Pierre
Competitive Quality Strategies

POMEROL Jean-Charles
Decision-Making and Action

SZYLAR Christian
UCITS Handbook

2011

LESCA Nicolas
Environmental Scanning and Sustainable Development

LESCA Nicolas, LESCA Humbert
Weak Signals for Strategic Intelligence: Anticipation Tool for Managers

MERCIER-LAURENT Eunika
Innovation Ecosystems

2010

SZYLAR Christian
Risk Management under UCITS III/IV

2009

COHEN Corine
Business Intelligence

ZANINETTI Jean-Marc
Sustainable Development in the USA

2008

CORSI Patrick, DULIEU Mike
The Marketing of Technology Intensive Products and Services

DZEVER Sam, JAUSSAUD Jacques, ANDREOSSO Bernadette
Evolving Corporate Structures and Cultures in Asia: Impact of Globalization

2007

AMMI Chantal
Global Consumer Behavior

2006

BOUGHZALA Imed, ERMINE Jean-Louis
Trends in Enterprise Knowledge Management

CORSI Patrick *et al.*
Innovation Engineering: the Power of Intangible Networks

Printed in the USA
CPSIA information can be obtained
at www.ICGtesting.com
LVHW020745180124
769123LV00005B/186